FORGOTTEN LANDSCAPES

ALSO BY STANLEY A. RICE

Encyclopedia of Evolution
Green Planet: How Plants Keep the Earth Alive
Life of Earth: Portrait of a Beautiful, Middle-Aged, Stressed-Out World
Encyclopedia of Biodiversity
*Scientifically Thinking: How to Liberate Your Mind, Solve the World's Problems,
    and Embrace the Beauty of Science*

FORGOTTEN LANDSCAPES

How Native Americans Created
Pre-Columbian North America
and What We Can Learn From It

STANLEY A. RICE

Essex, Connecticut

PB Prometheus Books

An imprint of The Globe Pequot Publishing Group, Inc.
64 South Main Street
Essex, CT 06426
www.globepequot.com

Distributed by NATIONAL BOOK NETWORK

Copyright © 2025 by Stanley A. Rice

All rights reserved. No part of this book may be reproduced in any form or by any electronic or mechanical means, including information storage and retrieval systems, without written permission from the publisher, except by a reviewer who may quote passages in a review.

British Library Cataloguing in Publication Information Available

Library of Congress Cataloging-in-Publication Data

Names: Rice, Stanley A., 1957– author
Title: Forgotten landscapes : how Native Americans created pre-Columbian
  North America and what we can learn from it / Stanley A. Rice.
Description: Essex, Connecticut : Prometheus, 2025. | Includes bibliographical
  references.
Identifiers: LCCN 2024061672 (print) | LCCN 2024061673 (ebook) | ISBN
  9781493088669 cloth | ISBN 9781493088676 ebook
Subjects: LCSH: Indians of North America—Antiquities | Indians of North
  America—Agriculture | Traditional ecological knowledge—North America |
  Land settlement patterns, Prehistoric—North America | Landscape ecology—
  North America | Social archaeology—North America | BISAC: HISTORY
  / Indigenous Peoples in the Americas | SCIENCE / Philosophy & Social
  Aspects
Classification: LCC E77.9 .R54 2025 (print) | LCC E77.9 (ebook) | DDC
  970.004/97—dc23/eng/20250327
LC record available at https://lccn.loc.gov/2024061672
LC ebook record available at https://lccn.loc.gov/2024061673

♾ ™ The paper used in this publication meets the minimum requirements of American National Standard for Information Sciences—Permanence of Paper for Printed Library Materials, ANSI/NISO Z39.48-1992.

CONTENTS

Introduction: The View from Mantle Rock	1
1 Welcome to America: Before 1492	9
2 Forgotten Fires: How Natives Used Fire to Manage the Landscape	33
3 Masters of the Hunt: How Native Hunting and Fishing Controlled Animal Populations	61
4 Farm or Forest? Agriculture in Prehistoric America	79
5 The Blessings of Water: Irrigation in Prehistoric America	99
6 Little Paradises: Orchards in Prehistoric Native America	121
7 European Diseases: The Beginning of the End of Native America	143
8 A Toxic Paradise: The San Joaquin Valley as the Ultimate White Monoculture Dream	161
9 Toward a Healthier World, with a Little Help from Your Local Natives	173
Epilogue: What We Have Lost, and What We Can Regain	183
Appendix	201
Notes	209
Dedication and Acknowledgments	227
Index	229

INTRODUCTION

The View from Mantle Rock

My great-great-great-grandmother Elizabeth Hilderbrand Pettit walked the Cherokee Trail of Tears in 1838.[1] She was the great-granddaughter of the famous Cherokee leader Nanyehi, who became Nancy Ward. Nancy Ward worked tirelessly for peace between the Cherokees and the white invaders (Europeans, and later, Americans).

Despite Cherokee efforts at peace, from Nancy Ward until John Ross, the chief during the time of the Trail of Tears, the American government forced the entire Cherokee tribe from their prosperous homes in the Appalachians and sent them west to land that later became Oklahoma. The government proclaimed that the Cherokees were savages and were not using the land for agriculture and civilization the way God intended it to be used, the way the descendants of European settlers were using it. About halfway through the journey, at Mantle Rock in extreme northwest Kentucky, the Cherokees and their American army escorts had to stop for a long time until they could cross the icy Ohio River.

While camped at Mantle Rock, Elizabeth had time to look east toward her homeland and to reflect on the events that had led up to the Trail of Tears. Even before their first contact with Europeans about a century earlier, the Cherokees were not savages. Their society was based on agriculture and settled villages. The Europeans, and later, the Americans, knew this only too well, because in order to conquer the Cherokees, English (then American) soldiers had to pull up cornstalks from their fields one by one in order to starve them out, something that would have made no sense if the Cherokees were hunters and gatherers.

Next, Elizabeth looked west into a future she could not imagine, into what Cherokees called the Darkening Land, where spirits go when they

2   *Forgotten Landscapes*

die. She was at the cusp between two worlds—the old Native American world, and the new world dominated by white Americans.

The Cherokees, as well as other Native Americans, were not savages, running through the trees, gathering nuts and berries, leaving hardly a footprint behind them. They made intense and profitable use of the landscape. This book is about the largely forgotten landscape that Native Americans created before European contact.

Starting in 1492, Europeans thought they had found a wilderness, untouched by human hands. But North America was not a wilderness. When the Europeans arrived in North America, a great Native civilization of clean cities and healthy people had just collapsed, leaving behind mysterious burial mounds and a continent-wide trade network (chapter 1). With the controlled use of fire (chapter 2), Native Americans had transformed thick forests into open woodlands and expanded the ranges of prairies. Through organized hunting, the Natives controlled the populations of prey animals such as passenger pigeons (chapter 3). When Native populations grew large enough, they developed agriculture (chapter 4), including irrigated crops (chapter 5), and even orchards (chapter 6). From the ruins of the collapsed civilization, Natives developed a system of government from which the Founding Fathers derived some of their ideas of democracy. The Native transformation of North America into a productive landscape is what this book is about.

The Europeans, and later, Americans, looked directly at the results of Native fires, farms, and orchards and didn't recognize them as being anything other than the effects of nature. They looked at earthwork mounds left by earlier civilizations and saw nothing but little hills. The white pioneers were probably unaware that the very trails on which they walked and rode had been established and maintained by Natives for many centuries as a continent-wide trade network.

We will probably never know what North America, without Native impact, would have been like. About fourteen thousand years ago, the glaciers of the most recent ice age began to melt, revealing new soil, on which trees, grasses, and shrubs began to grow. As the glaciers retreated northward, the land itself, freed from its burden of ice, began to spring back to higher elevations, and sea level began to rise from the infusion of glacial meltwater. Bands of vegetation, and the ranges of animals, moved northward. Tundra, the coldest vegetation zone in the world, used to be found in Kansas, but after a few thousand years of global warming it retreated to northern Canada. By about five thousand years ago, the vegetation of North America was approximately what it is today.

Introduction   3

But even as the glaciers retreated, the ancestors of modern Native Americans (sometimes called Paleoindians) entered a land truly unpopulated by other humans. Right away, they began setting fires and hunting. Both natural and Paleoindian fires occurred, but at the same time and in the same places. This impact was profound enough that, even now, scientists are unsure to what extent the prairies of central North America were the product of natural fires and to what extent they were the product of fires deliberately set by Native Americans. There is no wilderness against which to compare the effects of Native American activities. All we can do, as I do in this book, is to examine the scientific, archaeological, and historical evidence and try to reconstruct the surprising past.

IN ELIZABETH'S FOOTSTEPS

The day I visited Mantle Rock, it was breathlessly hot and humid. Far away from the nearest large city, Mantle Rock is a natural sandstone arch. It is surrounded by a forest of sugar maple, shagbark hickory, American elm, and hophornbeam. I looked around at the intense greenery and thought about how beautiful a place this was (see figure I.1 on next page). It is less than a mile away from the Ohio River.

It was very different when, 181 years earlier, Elizabeth Hilderbrand Pettit was there. The United States government forced as many Cherokees as they could find, even the ones hiding back in the hollows of the Appalachian forests, to leave everything behind so that white Americans could take over their farms, livestock, orchards, and houses. The year 1838 had a ferociously cold winter. Over a thousand Cherokees in the Hilderbrand contingent—one of the dozen contingents on the Trail, this one led by Elizabeth's cousin Peter—camped in the snow and ice. The Rock itself provided little shelter; most of the Cherokees camped out under the bare branches. Many Cherokees were ill and died at this place.

The Cherokee Nation in the Nineteenth Century

The federal government kicked the Cherokees off of their land because white Americans desired it. But they needed, or at least wanted, an excuse to justify this action. The excuse they gave was that the savage Native Americans were not using the land to its full potential.

But Elizabeth knew this was a lie. She knew that the landscape of eastern North America was not just a spot of earth on which her tribe *lived*

Figure I.1. Mantle Rock, Kentucky, where the Hilderbrand contingent of the Cherokee Trail of Tears stopped in the winter of 1838. *Author photo.*

but was a landscape her tribe helped to *create*. She was neither hunter nor gatherer; she herself had a farm and livestock in the Cherokee Nation.

In the nineteenth century, the Cherokees still raised corn as they had in previous centuries, though they now used plows and oxen rather than digging sticks to prepare their fields. Beef and pork replaced venison as their source of meat. The Cherokees even made themselves as white as possible in terms of their society and social economy. In place of traditional thatch-roofed longhouses, they now had white-style houses with porches and barns. They had long before started using horses, which came from Europe, both to ride and to draw their carriages. They not only learned to write, but the Cherokee leader Sequoyah developed a Cherokee writing system. Many Cherokees had adopted Christianity, and the Nation was full of churches, many of them with Cherokee preachers. The Nation even had a constitutional government modeled after that of the United States, including a Cherokee Supreme Court. As Cherokee leader John Ridge said in 1832, addressing the US federal government,

> You asked us to throw off the hunter and warrior state. We did so— you asked us to form a republican government: we did so—adopting your own as a model. You asked us to cultivate the earth, and learn the mechanic arts. We did so. You asked us to learn to read: we did so. You asked us to cast away our idols, and worship your God. We did so.

White settlers saw them do all of these things. To believe that Natives were wild savages, the white settlers had to disbelieve what they saw with their own eyes.

It almost seems as if the Cherokees were saying, *We have white farms, white livestock, white houses, white churches, even a white government. What more could you want? Can't you just leave us alone?*

And there was yet another way in which the Cherokees emulated their white neighbors. The Cherokee Nation had a long history of including women in their decision-making processes. Perhaps Elizabeth's thoughts could best be summarized in the words that Becky Hobbs and Nick Sweet put into their musical *Nanyehi* over two hundred years later: "Cherokee women have always done and will always do what they want." But the new 1827 Cherokee constitution excluded women from the government. Nancy Ward was the last female leader in the tribe by the time of her death in 1822. In effect, the Cherokee Nation was saying to the white Americans, *We even disenfranchised our women, just like you. What more could you want? Won't you just leave us alone?*[2]

6   *Forgotten Landscapes*

The answer was no, so long as the whites did not have Cherokee land. All of it. Through a series of treaties, Cherokees had been forced to give up large parcels of their homeland. The white Americans, particularly the residents of Georgia, wanted the rest of it. And not just the land, but also the gold mines of Dahlonega. The lust for gold that would eventually destroy the California Natives and the Sioux tribes of the Black Hills got an early start in the Cherokee Nation. The result was the Trail of Tears.

The Cherokee Nation before the Nineteenth Century

Elizabeth probably knew a little bit about the history of the North American continent. She knew that the whites had come from Europe, far across the ocean, from a land Nancy Ward's uncle had visited in 1730. She knew that the Europeans had carried out atrocities against the Natives, starting with Christopher Columbus. She also knew that the Europeans had brought many deadly diseases, which caused repeated waves of epidemic illness among the Cherokees, including the 1738 smallpox epidemic which had scarred the face and the personality of Nancy Ward's warrior cousin Tsiyu Gansini (Dragging Canoe). Before the white immigrants came, plagues were nearly unknown in America (chapter 7).[3] And Elizabeth knew that, before the Europeans, the Native tribes had sustained themselves by planting crops.

Elizabeth could not have known that, about six hundred years before her time, a large Native civilization had flourished along what is now the Mississippi River drainage system, a civilization even the Natives had forgotten about, a civilization with a large-enough population that it could not have survived on hunting and gathering. Elizabeth would have known even less about the Native tribes who migrated from Siberia about fifteen thousand years ago, something which in her mind was concealed behind a screen of mythology. Perhaps at that time the wild-man image of Natives was true, just as it was for European tribes at the same time. Elizabeth knew that the Cherokees were not savages and had not been savages for a very long time.

THE DARKENING LAND

When Elizabeth arrived in what would later be Indian Territory and then Oklahoma, she received a plot of land and enough money and provisions to start a new farm pretty much like her old one. For Elizabeth herself, the

Trail led to her old life in a new place. Cherokee cattle would replace the bison hunted by the Caddo and Comanche, who already lived in the territory chosen by the United States government as the new Cherokee homeland.

Elizabeth could not imagine the white transformation of the landscape. The white impact included the deliberate near extinction of the bison in the late 1800s; enough soil erosion that Oklahoma dust would darken the sky in Washington, DC, in the 1930s; and the industrial pollution and pesticides of the late twentieth century. She could not imagine that she would have a great-great-great-grandson who would grow up in a California valley that looked like an agrarian paradise but was an artificial and toxic one (chapter 8). And she could not imagine that there would even be such people as environmentalists. What we call environmentalism was something that, to Natives, was just the most sensible way of making a sustainable living (chapter 9). Elizabeth could not have imagined that people in the twentieth and twenty-first centuries would look to Native tribes for ideas about how to save the land, looking right past the benefits of Native land management to see a vague spirituality based largely on the gathering and use of Indian medicinal plants.

It is impossible for me to write, or for you to read, a comprehensive book about Native Americans and the land. I have written mostly about the Cherokees, because that is my tribe and I know a lot about it from my family history. And I know about Native agriculture and fire, and their ecological impacts, as a scientist. But I know that other tribes have had similar experiences. All Native tribes suffered conquest, and the other Oklahoma Native tribes had their own Trails of Tears.

We can never go back to the way the continent was even when the Natives managed it. But we can, by making use of Native insights, harm the landscape a little less and, maybe, create a sustainable future. It is my hope that this book will help the United States turn away from its environmental devastation and give the successful patterns of Native land management, as described in this book, a try.

1

WELCOME TO AMERICA

Before 1492

When Christopher Columbus "sailed the ocean blue," the continent we today call North America (north of Mexico) just might have been the best place in the world to live. Especially, as I mentioned in the introduction, if you were a woman.

Pre-Columbian North America was not a utopia, although Sir Thomas More might have based the original 1516 "Utopia" on fragmented stories he had heard from explorers returning from North America.[1] But compared to every other place on Earth, North America was healthy and prosperous.

Rutger Bregman, a Dutch author, wrote *Utopia for Realists* in 2017.[2] One of his basic points was that Earth, all of it, is a better place for almost everyone to live than it was just two centuries ago. Even people living at the poverty line in the United States are better off than over 80 percent of the people in the rest of the world. Bregman points out that in 1820, 84 percent of people lived in what we would recognize as extreme poverty. By 1981, this was reduced to 44 percent, and by 2015, to just 10 percent.

Bregman makes a broad, sweeping generalization that blinds us to a great deal of important information. He said that for 99 percent of history, 99 percent of humanity was "poor, hungry, dirty, afraid, stupid, sick, and ugly." He wrote that "the lives of almost everyone almost everywhere have almost always been grim." He made a vivid contrast between modern and medieval Europe. For example, the murder rate in Western Europe today is forty times lower than it was in medieval times.

This is true if one compares modern with medieval Europe. But this comparison does not hold true for pre-Columbian North America. While medieval Europe must have been grim, at that time North America was not. Most people in the world now live better lives than anyone did in

10 *Forgotten Landscapes*

pre-contact North America, but the comparison I wish to make in this chapter is between medieval Europe and North America at that same time.

At the time of European contact, North America was more densely populated than most people realize. A centralized civilization, now called the Mississippian, rivaled any in Europe or Asia. It had just collapsed, though most of its economic activities (such as agriculture and trade) continued after the collapse. During the Mississippian civilization, people and their environment were healthy. The cities were relatively clean. The inhabitants were probably well-fed, strong, and had few diseases. There was very little poverty. Even the captives from rival tribes had a better life than the average European peasant, and many captives were adopted into the tribes as citizens. The people lived in freedom, at least more freedom than could be found anywhere else in the world. In particular, women suffered far less oppression in North America than in the Old World. Their opinions, though secondary to those of men, were taken far more seriously in the Native American councils of power than probably anywhere else in the world.

The many separate tribes, each with their own language, seemed to be constantly at war with one another. Most Native men, at least part of the time, were warriors. In Mexico, the Aztec Empire was exquisitely cruel. The priests offered almost daily human sacrifices (of captives from other tribes) to their sun-god. The European conquistador Hernán Cortés was disgusted by the bloodshed that he saw, and it took a lot of bloodshed to disgust a conquistador. It is possible that the Mississippian civilization north of Mexico was just as cruel as that of the Aztecs, though archaeologists have found comparatively few skeletons of Mississippian human sacrificial victims. At any rate, the centralized Mississippian empire had just been replaced by a strong and sustainable confederacy of large villages. Warfare was constantly in the background of Native American life, except when it was in the foreground. But the same thing can be said about Europe at that time.

THE BEGINNING

About fifteen thousand years ago, according to most historians, the ancestors of Native Americans migrated out of what is today Siberia.[3] Or it may have been earlier. The 2023 discovery of twenty-thousand-year-old fossilized footprints in New Mexico may cause us to revise our historical timeline.[4]

During the course of about a half millennium, the First Americans filled every habitat from forest to desert, from seashores to the highest mountains on the two continents, right to the chilly tip of South America. They had no goal in mind, but merely moved to new places that had more game and plant foods. There were separate and perhaps earlier waves of immigration from Siberia, mostly along the Pacific coast, leaving artifacts at such places as Monte Verde[5] in Chile and on California's Channel Islands.[6]

The "Paleoindians," in tribes now forgotten, left arrowheads and spear points with a distinctive design, called Clovis tips, and carved a few petroglyphs. But their populations were small and their impact on the land was minor. Some scholars have claimed that the Paleoindians drove numerous species of large mammals—such as mammoths, mastodons, and giant ground sloths—to extinction by overhunting (chapter 3). This "Pleistocene extinction," however, occurred at a time when Native populations were still small.[7] Overhunting by Paleoindians may have contributed to, but is unlikely to have been the main cause of, the Pleistocene extinction, which occurred throughout the Northern Hemisphere, not just in North and South America.

As Native populations grew, they began to have a significant effect on the North American landscape. At first, it was mostly by the controlled use of fire (chapter 2) and by communal hunting (chapter 3). Humans had used fires in their hearths for at least three-quarters of a million years, long before the evolution of modern *Homo sapiens*. But the newcomer *Homo sapiens* in North America did not just cook dinner but set fires to, and transformed, entire habitats.

As their populations grew even more, Native Americans developed agriculture (chapter 4). The Natives in Mexico bred what are some of the most important crops in the world today, including maize, certain types of beans, squash, tomatoes, chili peppers, and many others. The Natives of highland South America domesticated potatoes. In this way, Native Americans had a major and lasting worldwide impact on modern agriculture. At first they did not have agricultural fields, but they had gardens to supplement their hunting of wild game. And where there was not enough water, they dug canals (chapter 5). They also maintained fruit and nut orchards (chapter 6).

Sometime before 1100 CE, the population of Natives reached a critical mass, and their villages, already connected by strong networks of trade, began to amalgamate into an empire supported by large-scale agriculture: the Mississippian culture.

12   *Forgotten Landscapes*

WELCOME TO CAHOKIA

American tourists spend a lot of money to visit medieval European and Asian sites, not realizing that there was a great medieval civilization right here in North America. Even travelers who like to visit American historical landmarks usually go to Revolutionary and Civil War sites. If they are interested in Native American history, they are likely to visit places known for battles between Natives and whites, such as Sand Creek or Little Bighorn. The Mississippian civilization, whose cities once rivaled those of Europe, has been largely erased from the landscape and from memory.

When Europeans first became aware of the Aztecs of Mexico and the Incas of Ecuador and Peru, they could hardly believe that the reddish-skinned pagan Natives could have produced these civilizations themselves. Yet there they were, with huge pyramids and large buildings built of carefully worked stone. And with lots of gold and silver. To Spanish eyes, these Native American civilizations were ripe for conversion and conquest.

It was clear to the conquistadors that these civilizations had antecedents. The Mayas dominated Central America long before the Aztecs; their temples and even their astronomical observatories were still visible under the jungle vines of Yucatan and Central America. The Moche and other civilizations that preceded the Incas were also evident in abandoned cities such as Tiwanaku on the Altiplano of Bolivia. Many Europeans imagined that these predecessors of the Aztec and Inca civilizations must have been immigrants from the ancient Old World civilizations such as the Egyptians and Babylonians. Some scholars imagined that Native languages were similar to and derived from ancient Middle Eastern languages. Some even believed that the Native Americans were the remnants of the Ten Lost Tribes of Israel, even though there is no clear evidence—even from the Bible—that these ten tribes were in fact lost. Amateur historians even imagined that Aztec pyramids were imperfect copies of Egyptian pyramids.

The evidence for the independent evolution of Central and South American civilizations soon became overwhelming. By the time Gregory Mason wrote his popular book *Columbus Came Late* in 1931,[8] most people knew that the Mayans, Aztecs, and Incas were completely American civilizations, even though some popular books, including one from a major publisher as recently as 2020, still claim that Native American civilizations were derived from the Old World.

The apparent similarities between Native American and Old World civilizations were due to convergence—that is, the independent evolution of social structures in response to the same needs and opportunities, and based on the same human nature. For example, Egyptian and Aztec pyramids may look alike at first glance. But the Egyptian pyramids were burial structures, not intended to be climbed. Mayan pyramids, in contrast, were also burial structures but had steps that led to ceremonial sites at the top. They, like the Egyptian pyramids, had four sides, but this could be a human instinct; the four directions are forward, backward, right, and left, and probably would be so for any bilaterally symmetrical motile animal in the universe.

Nor is it surprising that the alignments of megalithic structures would follow astronomical coordinates. The Natives of America, like the people of the Old World, would have noticed and been intensely interested in the apparent movement of the sun from its lowest to its highest arc and back, and the dates on which the sun appeared to retreat and to return. Such apparent solar movements would give people advance, and reliable, indications of the changes in the seasons. They could order their lives around the movements of the sun. Native Americans, like the people of the Old World, had sufficient intelligence and motivation to figure this out. The Native Americans developed their own astronomical knowledge.

The Mayan and Tiwanaku civilizations collapsed because their land use and tribal wars were unsustainable, as explained in Jared Diamond's book *Collapse*.[9] The Mayan collapse was caused by the intersection of an unsustainable civilization and a drought, neither of which by itself might have brought about the end of Mayan civilization.

The Mississippian culture of North America also had large cities. Cahokia, located in what is today southern Illinois, was perhaps the largest. The biggest mound in Cahokia—known as Monk's Mound, because a couple of centuries ago there was a monastery on top of it—commands a view of the surrounding countryside of forests and grasslands. To the west, on the Missouri side of the Mississippi River, you can see St. Louis. Aside from an interpretive center, there is hardly anything at Cahokia today. Concrete stairs allow visitors to walk, or occasionally run, to the top of Monk's Mound (figure 1.1).

That's about all there is today. When I visited, I tried to imagine what it was like nine hundred years ago, at the peak of Mississippian civilization. At that time Cahokia may have had twenty thousand inhabitants, as many people as London had at the same time, and larger than Philadelphia's population during the American Revolution. Centuries ago, from the largest mound, you could see many other earthen mounds, inside of which bodies of important people and luxury artifacts were buried. On platforms at the tops of the mounds, warrior-priests carried out religious ceremonies,

Figure 1.1. Today, Cahokia's mounds are preserved by the state of Illinois. Hundreds of people climb Monk's Mound, with the aid of a concrete staircase that was not present during the Mississippian era (top). From there, one can see numerous smaller mounds (bottom). *Author photos.*

Welcome to America   15

through which they kept countless thousands of people spiritually enthralled to them, the way churches in Europe were doing at the same time.

There would have been the bustle of many thousands of people at Cahokia. In workshops all around the mounds, craftsmen made ornaments and tools. Archaeological analyses have shown that many of the artifacts found in other city-mounds in North America were made here. The artifacts that were not actually made on-site were imported along the networks of trails that spread throughout the continent, as far as you could see in any direction you cared to look. Artifacts were made from shells from the Gulf of Mexico, copper from the Great Lakes region, and soft stone used for pipe bowls from regions such as what is today southwestern Minnesota. Cahokia was also the hub of trade for log barges going up and down the Mississippi River, just as St. Louis is today a hub for ship traffic. The river is now hidden by trees, but nine hundred years ago you could see the river, because the Natives managed the landscape with fire, in addition to having extensive agricultural fields.

With all the religious splendor and manufacturing activity, Cahokia might have resembled a major European city. Many square miles of farmland—mainly maize, beans, and squash—spread all around, to feed not only the thousands of residents but the thousands of visitors who came for religion and trade. These throngs of people were in a hurry. The wealthy wore colorful costumes and jewelry; the less wealthy wore plain clothes of fiber and leather.

Shouts of pleasure would have mixed with screams of agony as the warrior-priests ordered mass executions of captives. In a pit in Mound 72 at Cahokia, numerous young women were sacrificed, and four young men were killed, their elbows linked, and their heads and hands cut off. This suggests that Cahokia was not just a big city but the capital of a brutal empire. Evidence of Mississippian human sacrifice, however, is rare.[10]

No one knows how far Cahokia's sphere of military influence extended, but its sphere of economic influence included most of eastern North America. Unlike any of the famous empires of Mexico, South America, Europe, Asia, or Africa, the physical structures of the Mississippian culture have nearly vanished from geography—and from memory. They are now just mounds of earth.

As I noted previously, most of the food came from crops, especially maize. But the inhabitants also ate meat. For this, they hunted deer and other mammals, and caught fish (chapter 3). Natives had domesticated turkeys. North America did not have native species of mammals that could be domesticated, with one exception: dogs. Dogs were mainly

16   *Forgotten Landscapes*

used to transport items on travois, which are little platforms which the dogs dragged in the dirt, and for defense. There were no wild cattle. North American sheep and goats were not the passive grazers that were found in Eurasia, but rather wild animals that scampered on the cliffs. Deer, elk, and their relatives have never been domesticated. An animal species suitable for domestication has to be willing to follow the herd, and to be enclosed. Enclosing a deer? The Natives would have had as much success with this as did the little boy in Marjorie Kinnan Rawlings's *The Yearling*.

Cahokia is very different today. What was once the richest place on the continent is now one of the poorest. Novelist Richard Powers has quipped that it's visited only by schoolchildren under duress. It is right next to the economically depressed East St. Louis. I stayed at just about the only motel in the vicinity and had to eat at the local convenience store, where I watched homeless people sifting through the garbage.

WELCOME TO SPIRO

Spiro, in what is now Oklahoma, was almost as large as Cahokia at the same time, but probably had a smaller resident population. Its people seem to have been mostly pilgrims who came for religious observances. But, like Cahokia, it had mounds in which artifacts, and bodies, were buried. Like Cahokia, Spiro was on a major waterway, the Arkansas River, accessible from numerous trade networks.

Spiro is today a little Oklahoma town on a bend of the Arkansas River, surrounded by acres of hayfields. It is not on the road to anywhere at all. To the north and east is the Arkansas River. The only way out of Spiro is the way you came in. Almost nobody visits the Spiro archaeological site, even those who live just a few miles away. I got lost looking for it and had to ask for directions. We finally had to ask Siri how to get there.

I eventually found the Spiro Mounds Archaeological Center. The day I was there, I was almost the only visitor. I stood near a mound of earth, partially covered with trees and shrubs, about fifteen feet above the surrounding plains. There were other mounds nearby, making a total of twelve, all but one of which are now just little swales of earth. Spiro Mounds looks just like any other pasture and woodland in the eastern Oklahoma countryside.

These little mounds are all that remain of what was, around the year 1100, perhaps the largest human-built structures in North America north of Mexico, maybe even larger than those at Cahokia. Like Cahokia, Spiro was one of the major cities of the world, equal in size and importance to many famous European cities. It was a totally Indigenous civilization, unrelated to anything in Europe or Asia, and with probably very little connection (aside from the crops) to the Mayans who lived at the same time.[11]

In addition to Cahokia and Spiro, there were perhaps hundreds of other cities and mounds throughout what is now Ohio, Wisconsin, Oklahoma, Missouri, Louisiana, and Georgia. One of them is Etowah Mound near Cartersville, Georgia (figure 1.2). At least sixty sites have been explored, but others may remain undiscovered. In 2023, remains of a Mississippian-era village were found in Lake Mendota, in Madison, Wisconsin, though whether it was a metropolis, as its discoverers claim, remains to be seen.

Figure 1.2. Etowah Mound, near Cartersville, Georgia, was one of the secondary cities of the Mississippian Culture, similar to and much better preserved than Spiro Mounds in Oklahoma. This mound, like Cahokia, has been reconstructed. *Author photo.*

18 *Forgotten Landscapes*

The Mississippian was a major civilization, built of perishable materials and without written records. The modern tribes descended from them (such as the Caddo) have no traditional knowledge of the Mound Builder culture of their own past. Scholars deduced the connection between the Mississippians and modern tribes by similarities in art forms. When botanist William Bartram traveled through much of what is now the southern United States in the eighteenth century, he saw many mounds, but the Natives told him they knew nothing about them.[12] Certainly by the time of European contact, nobody was building mounds or great cities in North America anymore.

TRADE NETWORKS

While the Mississippian civilization itself vanished before our year 1350, much of the economy did not. The Mississippian complex was gone, but much of its economic structure remained in the Iroquois and Powhatan confederacies in New York and Virginia, which were active when the first Europeans arrived. In particular, the advanced systems of agriculture that had supported millions of people in Mississippian times were still in use, though not supporting as many millions of people as before. The Europeans observed Native villages surrounded by wooden palisades and their agriculture, noting that many of the trade objects the Europeans got from the Natives had come from a great distance. In place of a civilization with great cities, the Europeans encountered trade networks of large villages. These trade networks may have been the strongest component of Mississippian civilization. They allowed that civilization to emerge, and they persisted after it fell.

For decades I hiked the trails of Turkey Mountain on the west side of the Arkansas River opposite Tulsa, Oklahoma. Only when I started writing this book, however, did I realize that these trails were exactly the sort that unified the North American Native tribes and villages for many centuries (figure 1.3). The dirt trails were like the ones on which messengers and traders traveled from one village to another during the pre-contact millennium, carrying leather bags of trade goods and wearing moccasins. The pre-Columbian Natives had no horses. All Native horses were descendants of the ones that the Spaniards brought to southwestern North America in the sixteenth century. Of course, the Natives had no wagons either, or oxen to pull them. They had wheels, but used them only on toys. I looked at the trail with new eyes, seeing my hikes as a continuation of the activities of my Native ancestors.

Figure 1.3. A modern trail on Turkey Mountain, west of Tulsa, Oklahoma, resembles trails that were part of the trading network of the Mississippian Culture. Due to its strategic placement, this trail might itself have been a Mississippian-era trail. *Author photo.*

20   *Forgotten Landscapes*

Since the trails followed the contours of the landscape, very few of them now remain. They have been covered by modern roads and highways, which follow the same paths as the original trails.

The Arkansas River was a major conduit of trade by Native barges. All the barges heading down to Spiro would have gone this way. There had to be trails on both sides of the Arkansas River to accommodate the flow of trade goods going east or west across the river. Turkey Mountain might be one place where you can still see the original trails. Modern roads run parallel to the Arkansas River, on both the east and west sides, but the road on the west side is about a mile away from Turkey Mountain, inconveniently far from the river for Mississippian-era foot traffic. There had to be a trail along the west side of the river from which the porters could get a good view of it.

I realized that some of the trails I hiked on Turkey Mountain—along the crest, and along the east-facing bank—might actually be the same ones the Mississippian Natives used for their trade networks. They may have been the same trails used by the Osage tribe, who occupied the area after the Mississippian civilization collapsed. They may have been the trails used by the Muskogee tribe, which was sent here by the United States federal government from their homeland in Georgia and Alabama in 1832, just before the Cherokee Trail of Tears.

None of the modern trails could be the direct, uninterrupted descendants of the pre-contact trails. During the 1920s, Turkey Mountain was an oil field. You can still see iron pipes, cement structures, and iron cables left over from petroleum extraction. In many cases, the corporations that profited from Tulsa oil extraction obtained their oil leases from the legitimate Native owners by kidnapping them and taking them out of the county, or out of the state, or in one case, even out of the country (a ploy used by a prominent Tulsa oil man in the early twentieth century) to get them to sign over their oil rights.[13] In some cases, the Native owners mysteriously died. The trail on which I hiked was not only historically drenched in oil but also in oppression and blood. When the remaining oil under Turkey Mountain was no longer profitable to extract, the corporation abandoned the operation. The City of Tulsa obtained it for what is now Turkey Mountain Urban Wilderness. (There are no turkeys; it is not big enough to be a mountain; and it is certainly not a wilderness.)

The trade networks were important not only in the forested eastern part of North America but also in the plains. Some tribes specialized in hunting, others in agriculture; then they would meet at major trade centers. The hunters would exchange extra meat for agricultural and manufactured goods. These trade centers included locations on the Missouri River (for the Mandan), the Republican and Platte Rivers (for the Pawnee), the Canadian and Red Rivers (for the Caddo), and the Rio Grande (for the Pueblo).[14] Each culture bartered what it had in order to obtain what it did not have.

CLEAN CITIES AND HEALTHY PEOPLE

Whatever else can be said of cities such as Cahokia and Spiro, we can be fairly certain that they were clean. While we have no proof of it, it is difficult to avoid this inference. Big cities of pre-Columbian North America were almost certainly not cesspools the way London was at the same time. If they were, the Natives, no matter how few diseases they brought with them from Siberia, would probably have developed their own diseases that the Europeans, foisting their dirty selves upon the Natives, would have taken back to Europe. As it was, it appears that the only disease the Europeans took back with them was syphilis, which is not associated with human wastes.

Clean Cities

Whenever humans build cities, and live in concentrated areas, they need to have some way to deal with wastes if the cities are to be sustainable and livable. Medieval European cities did not do this. Running-water toilets were widespread in Roman times and known as far back as Cretan times, but post-Roman Europeans generally saw no need for waste disposal. Homes sometimes had porcelain chamber pots. People would empty the chamber pots directly into gutters from their windows; this is supposedly the origin of the habit of the gentleman walking on the street side, so that he was more likely than the lady to get doused by carelessly emptied chamber pots. In some cases, people would stick their butts out of upper-story windows and relieve themselves directly into the street. Gutters were runnels of human waste. Some streets had names that reflected this ill fame—for example, Pyssynge Alley (now Chick Lane) and Shitebourne (now Sherbourne) Lane in London. Professional waste haulers were supposed to properly dispose of the wastes, but often did not. Human wastes mingled with animal wastes and blood from slaughterhouses.[15]

Sometimes street sweepers would clean away the ordure, and sometimes not. Public toilets were poorly maintained. In London, people counted on the tidal surges of the Thames to flush out the wastes. But

22   *Forgotten Landscapes*

frequently, waterways would get jammed with human wastes, which would then overflow during heavy rains. Some outhouses were just holes where the wastes would fall into a river or into the garden.

By 1420, London city officials decided something had to be done. They inspected all parts of the city and found waste problems everywhere they looked. Here is a partial report of one example (with partially modernized spelling):

> The common privy at Ludgate is defective and perilous and the ordure thereof rotteth the stone wall and maketh an orrible stench and a foul sight . . . it is a disgrace to the City that so foul a nuisance should be so often complained about but yet no remedy has been ordered . . . dung in the highway all through the year and for casting out hot piss that had stood under a horse for a month so that no man can pass there for the stink.

Perhaps none of this was very noticeable to people, many of whom never bathed, a habit lampooned by Mark Twain in *A Connecticut Yankee in King Arthur's Court.*

Dumping wastes inside the towns and cities is something that Europeans should have known to not do. There is a specific command in the book of Leviticus (chapter 23, verses 12 and 13) which says, "You must have a place outside the camp to go and relieve yourself. Take a digging tool so that when you relieve yourself you can dig a hole and cover up your excrement." On nature trails where outhouses are not available, the National Park Service instructs hikers to dig what are technically known as "cat-holes," in good Leviticus tradition. Most Europeans could not read or write, and most Bibles before the King James Version were in Latin, but at least the priests should have known the importance of public hygiene.

This accumulation of wastes in public places was not unique to Europe, but in other places it was less common. Marco Polo in the fourteenth century expressed his astonishment that the Chinese city now called Hangzhou had a series of canals that constantly carried the wastes away.

In Native American villages, people relieved themselves outside of the camp—in the case of the Cherokees, outside the palisade walls—in stark contrast to European practice at the time of first contact. Moreover, Native Americans had a tradition of bathing. Among the Cherokees, bathing was obligatory at least once a year, during the Green Corn Festival, when all litter in the streets and houses was swept away. In some cases, Natives took baths daily. When the Spaniards captured some Natives to present before the king and queen, the Natives were offended by the filthy conditions on board the ship and, as soon as they had a chance in Europe, they ran and jumped into a river to bathe.

Europe was pockmarked with reeking cities, while North America was like a breath of fresh air.

Healthy People

Not only that, but Native Americans were generally healthier than Europeans. One major reason for this was the more equitable distribution of wealth. Europe had teeming millions of poor people, undernourished and regularly devastated by plagues. I have read a lot of Native American history without ever encountering any reference to slums or to pre-Columbian plagues.[16]

Another reason is that Native Americans had a healthier diet. The result was that Native Americans avoided two of the major problems of European nutrition: scurvy and ergot.

Scurvy results from inadequate vitamin C. While nobody before the modern era of medicine and nutrition knew about vitamins, it was occasionally noticed that people who ate more fresh fruits and vegetables were healthier. Limes were particularly high in vitamin C. Since scurvy was a particular problem on long sea voyages, the British navy eventually required all sailors to eat preserved limes (leading to the nickname "limeys"). European peasants had some access to fresh vegetables such as turnips and carrots, which have a lot of vitamin C, but less so than Native Americans, who ate lots of vegetables and fruits.

Ergot poisoning was also something that plagued European peasants but not Native Americans. In both places, most people ate grains, but in North America the major grain was maize, while the peasants of Europe ate mostly bread, often made of rye. Rye plants were often infested with a fungus known as ergot (*Claviceps purpurea*). This fungus created reproductive structures that looked a lot like rye grains, but were darker. The fungus got harvested and replanted along with the rye, and ground with it into flour. Bread made from this infected rye flour contained a poison, lysergic acid diethylamide (LSD), which was not inactivated by the baking process. Thus, it was common for rye bread to have low doses of this poison.

People who ate this bread were often ill. The poison would constrict blood flow to extremities and would affect the nervous system. Symptoms of ergotism included the death of extremities such as fingers and toes, as well as mild to extreme convulsions and hallucinations. The fungus grew better in areas with cool, wet growing seasons, such as Central Europe.

24   *Forgotten Landscapes*

They were less common in areas where the winters were cold enough, such as Scandinavia, or the summer dry and hot enough, such as Italy, to restrict the growth of the fungus.

People understood that rye flour was less healthy than wheat flour, which only rich people could afford. Ergotism made its human victims more susceptible to many other diseases, both low-level chronic diseases and occasional outbreaks of severe disease. No equivalent type of food poisoning is known for maize.[17]

Sometimes ergot outbreaks could have severe effects. Many historians have linked ergotism to witch and werewolf crazes. It was usually poor people in Central Europe, who ate a lot of rye bread, who accused one another of being—or believed themselves to be—witches and werewolves, and outbreaks of these crazes often occurred after cold growing seasons. These resulted from hallucinations brought on by consuming LSD. This never happened with Native Americans, although it occasionally showed up among European immigrants to America, as in the witch hunts of the late seventeenth century in Salem, Massachusetts. But even apart from outbreaks of hallucinations, European peasants suffered the chronic effects of low-level ergotism.

Good nutrition also calls for the right balance of amino acids in protein. The right balance requires proteins from both grains and legumes, such as beans and peas. Native Americans had access to more legumes in their food, although beans and peas were not uncommon in the diets of European peasants. Europeans probably had a protein balance not very different from that of Native Americans.

Because they were cleaner and had more—and healthier—food than Europeans, Native Americans were undoubtedly better fed, bigger, and stronger than the European invaders.

WELCOME TO HEAVENER

The voyage of Christopher Columbus began an uninterrupted assault by Europe on North America. There were earlier contacts between Europe and North America, but they were all temporary. The most famous pre-Columbian contact was between Native Americans and Viking armies.

Wherever they went, the Vikings cruelly slaughtered any people they found. When they came to what they did not know at the time was North America, a land they called Vinland (now Newfoundland), they referred to the Natives as *skraelings*, which roughly translates as "miserable people." The Viking warrior attitude was captured in a T. C. Boyle short story ("We Are Norsemen"), in which Vikings slaughtered Native Americans just to put them out of their misery. Nonetheless, the Vikings encountered resistance in North America that they could not overcome, and they eventually retreated.

Compared to their failure in America, the Vikings were fairly successful in Europe, despite Europe's technological superiority to America. Vikings conquered part of what is now England and Ireland, producing a territory known as the Danelaw. They conquered part of France, now known as Normandy. They even took their boats on major rivers in what is now Russia.

One possible reason for the nearly unique Viking failure in North America is that they were sick and small, like other Europeans, while the Native Americans were healthy and strong. Nearly every illustration of Vikings shows them as muscular and tall. This is undoubtedly wrong. When Richard Fleischer directed the 1958 movie *The Vikings*, starring Kirk Douglas and Tony Curtis, he wanted to model the Viking ship after an actual archaeological specimen. The actors who played the warrior roles, however, could not fit into the ship. The set designers had to rebuild the ship, leaving more legroom for the actors. The actors were men of ordinary modern build, resulting from ordinary modern nutrition and exercise. Native Americans were also free of European diseases. Against the strong, healthy Native Americans, the miserable Vikings didn't stand a chance.[18]

Heavener, like Spiro, is a little Oklahoma town on the Arkansas River and is not on the road to anywhere. The main thing this town can boast about is a rock which bears carvings that say, in the Old Norse runic alphabet, *Glome Dahl*, or "Glome Valley." Many local people, as well as aficionados of alternative history, believe this carving was actually made by a Viking long before Columbus. Most scholars have concluded that the runic carvings were made by a Swedish immigrant in the nineteenth century. If so, it would be one of many fake runic carvings in northeastern Oklahoma, one of which is at an undisclosed location on Turkey Mountain.

The faculty lounge at Southeastern Oklahoma State University has a painting on the wall. It shows Native Americans agitated at the approach of a Viking warship coming down the Arkansas River. If any Vikings made it to what is now Oklahoma, that is *not* how they did it. If a Viking army had tried to fight its way down into what is now Oklahoma, they would have left a trail of bones and artifacts, none of which have been found.

26   *Forgotten Landscapes*

While the average Viking warrior was cruel and bloodthirsty, there must have been a few exceptions—individuals who, at least in their private thoughts, wondered if the people they tried to conquer were really as miserable as their leaders said. It is possible that when one or a few Vikings looked at the *skraelings*, they saw generally happy, healthy people with plenty of wealth to accommodate their needs. They saw productive gardens, and nobody was starving. It looked like a pretty good life. This same idea was reinforced later when Natives kidnapped European children, who often did not want to go back to their "civilized" way of life. Examples include Eunice Williams in Deerfield, Massachusetts,[19] and Cynthia Ann Parker in Texas.

But what if one of these Vikings decided to run away from the others and spend his life with the *skraelings*? This would have been very easy to do. After getting away from the other Vikings, he could have left town in a couple of hours by traveling on one of the numerous trails used for intertribal trade. The Natives might have welcomed the Viking escapee, who might have provided valuable inside information that helped them to eventually expel the Norse invaders.

Take *that*, Leif Eriksson!

Despite occasional discoveries that suggest a genetic input from Europe in Native Americans,[20] the overwhelming evidence is that Native American genes came from Siberia during a relatively brief immigration event about eleven thousand years ago. Any Viking or any other white genetic input was so slight as to be undetectable among modern Native Americans.[21]

But that doesn't mean it *never* happened.

A runaway Viking would have been able to hitch a ride with one of the thousands of traders who united the Mississippian cities from Minnesota to Louisiana, from Pennsylvania to Georgia. A Viking deserter could have traveled throughout the northwestern area of the Mississippian culture. Upon reaching the Arkansas River, he could have voyaged on one of the trade boats to the present location of Spiro, where he would have seen a great city. From there it would have been only a short trip to a secluded little valley south of the Arkansas River, near the present location of Heavener, Oklahoma, where he may have decided this was where he wanted to spend the rest of his life—in Glome Dahl.

My point is not to say that this actually happened, but that pre-contact North America was the kind of continent where a Viking could have comfortably traveled the two thousand miles from Vinland to Heavener without a struggle.

Welcome to America   27

WHAT HAPPENED TO THE MISSISSIPPIANS?

There was probably no single cause for the collapse of the Mississippian civilization. It is possible that the dense urban centers required intense agriculture that led to localized soil degradation. This would have happened at the same time as a continent-wide drought. Perhaps neither factor by itself would have caused agricultural and economic collapse, but the combination of them did. While the cities were no longer political or economic hubs, they were not totally abandoned, retaining some inhabitants.[22]

The Mississippian empire collapsed the same way the Roman Empire did. The central power (Rome, or Cahokia) lost power, and the peripheral cultures began taking care of their own affairs. The Mississippians apparently maintained a stronger trade network among their component cultures than did the former Roman colonies. But it is possible that, by the time of European contact, the erstwhile Mississippian culture resembled Europe of the late Middle Ages, each kingdom prospering on its own. The difference is that the Mississippians left no writings or stone monuments, and they, unlike the Romans, were forgotten.

While nobody is sure why the Mississippian civilization itself collapsed, one major reason must have been because the oppressive warrior-priests were too heavy a burden for their society and economy. Life went on, but now each village made its own decisions about what to do. The chiefs of each village got together to decide whether, and when, to go to war against another tribe, or against European invaders.

As a matter of fact, the Cherokees (and probably other tribes as well) had a separate peace chief and war chief for each large village. They recognized that the skill sets for waging war and for running the peace were different. This political arrangement was inefficient from the viewpoint of contemporary European civilization. It was certainly a source of endless irritation to the Europeans. Before military officers or representatives of European civilizations could get any tribe to agree to a partnership, they had to sit through endless speeches and ceremonies at which all the Native leaders, including women, got a chance to speak. Getting consensus under these conditions seemed impossible, but this is what the tribes did. It just took a long time.

And it sometimes didn't work. In the 1780s, the general consensus of the Cherokee chiefs was that they should not go to war against the new United States of America. But one of the most important chiefs thought that war was the only option. When Tsiyu Gansini did not get the consensus of

28   *Forgotten Landscapes*

the Cherokees for war, he went off to the vicinity of Lookout Mountain in Tennessee, next to the Georgia border, and formed his own Chickamauga communities. He fought the Americans until his death in 1792.

The Europeans could simply not understand this. They—all of them—were accustomed to having a king give the orders, and that was that. The French, for example, thought that Attakullakulla (Nancy Ward's uncle) was the king of the Cherokees, and they imprisoned him in Quebec in the 1740s. It took them a while to figure out that he was not the Cherokee king, that the Cherokees had no king, and that Attakullakulla could not order the Cherokees to do anything they did not want to do. Another Cherokee, Moytoy, was not an emperor until the Englishman Alexander Cuming said he was, before taking him on a grand tour of London, which was mainly a publicity stunt.

It is very likely that the Mississippians had absolute rulers. This was, apparently, a political failure. There are Cherokee legends of the Anikutani, who were powerful priests. Was this legend an unpleasant memory of Mississippian dictators? Spanish explorer Hernando de Soto described seeing kings and queens carried on litters, a dense population, and trade networks when he visited what was probably Cherokee territory about 1540. Was this one of the last vestiges of Mississippian power? The Natives learned from their experiences. From that point on, the Natives adopted a system of seeking consensus when making big decisions.

A few Native American leaders were taken to England to be shown to the public and introduced to Parliament and to the king. Here was a chance for the backward savages to learn about the grace and beauty of European civilization.

But that is not what happened. In 1730, several Cherokee leaders including Attakullakulla went to England and met the king. Then, in 1762, Lieutenant Henry Timberlake took a later generation of Cherokee leaders, including Ostenaco, on a ship to England. In both cases the Cherokee leaders created a positive sensation in England. Everybody, even the whores, wanted to meet them.

In 1730, Attakullakulla was not as impressed with the English as they were with him. It has been reported that when Attakullakulla visited Parliament, he asked his hosts where the women were. Cherokee leadership had long included the Women's Council, which was responsible for specified areas of decision-making. He did not understand how a society could ignore the viewpoints of the women.[23]

In 1762, before departing for England, Ostenaco gave a speech at William and Mary College in Virginia. One of the students who heard him was Thomas Jefferson, who would become the third president of the United States. Jefferson was predisposed to admire Ostenaco, because his family had cordial relations with Cherokees in Virginia. But Jefferson's impressions went far beyond ordinary admiration. He later wrote of Ostenaco's speech,

> I knew much of the great Outassete [Ostenaco], the warrior and orator of the Cherokee. He was always the guest of my father on his journeys to and from Williamsburg. I was in his camp when he made his great farewell oration to his people the evening before he departed for England. The moon was in full splendour, and to her he seemed to address himself in his prayers for his own safety on the voyage and that of his people during his absence. His sounding voice, distinct articulation, animated action, and the solemn silence of his people at their several fires, filled me with awe and veneration, although I did not understand a single word he uttered.

When the founders debated American independence, and later drafted a constitution, there was no European model for a democratic system of government. The best example to which they could look was the Iroquois Confederacy. Jefferson, as well as Benjamin Franklin, recognized that the kind of union of free states that the founders desired for America had already been created by the Natives. This fact impressed them much more than the writings of ancient Greeks and Enlightenment philosophers to whom credit for the idea of democracy is generally given. The European philosophers speculated that democracy was possible; the Native Americans showed that it could work. Franklin wrote a letter to James Parker objecting to the notion that Native tribes, such as the Six Nations of the Iroquois Confederacy, were ignorant savages:[24]

> It would be a strange thing if Six Nations of ignorant savages should be capable of forming a scheme for such an union, and be able to execute it in such a manner as that it has subsisted ages and appears indissoluble; and yet that a like union should be impracticable for ten or a dozen English colonies.

It may seem strange that the Native tribes, at the time of European contact, had little if any memory of who built the mounds. But just remember that the Mississippian civilization, with its abuses of power, might have been like a bad dream that everyone just wanted to forget. And they did.

30   *Forgotten Landscapes*

SPIRO: MORE THAN JUST FORGOTTEN

Nothing remains of the Cahokia and Etowah mounds except the mounds themselves. In Spiro, even the mounds have been erased from the landscape. Spiro Mounds didn't look like much more than small hills when white Americans discovered them. Only when obviously ornate Native artifacts kept turning up on the surface of farm and pasture did anyone near the little town of Spiro suspect it of being an archaeological site.

A group of amateur diggers formed the Pocola Mining Company in 1933 and got permission to plunder this archaeological site, paying the owner, who may have had no idea what was there, a total of $300. The Pocola miners then dug tunnels into the mounds and sold whatever they could find: pearls, hand-carved beads, necklaces, carved pipes, and other works of art made from copper, stone, wood, and pottery. They found and sold effigy pipes, T-shaped pipes, masks made of shell and cedar, arrowheads, spearpoints, axes, flint knives (some of them almost three feet long), and embossed copper plates and bowls. They found conch shells and ornaments, as large as a small modern plate, to be worn around the neck. The artifacts—that is, the ones we know about—are now found in museums around the world, including the Louvre.[25]

It is difficult to understand the business model of this company. They dug out the artifacts and sold them wherever they could and for whatever they could get. The miners sold many pieces for a few cents, or, for finer specimens, a few dollars, to curiosity-seekers at the mouth of the mine. Or they loaded artifacts into the trunks of their cars and sold them in Eastern towns. They did not keep a record of what they took or where they sold it. Whatever they could not sell, which included pottery, textiles, feathers, blankets, and fur, they discarded. They burned the wooden timbers that could have later been used, through the study of tree rings, to ascertain the dates of habitation. They broke up some of the larger artifacts to sell the fragments in order to make a little more money than they could get for the whole piece. And they threw out the interred corpses. Even during the Depression, this business model made no sense.

When it became apparent to state and federal officials that Spiro Mounds was a major archaeological site, the Oklahoma legislature passed an antiquities protection act. Until that time, the Pocola looting was legal in Oklahoma. But they had already violated federal law. As soon as any of the artifacts crossed state lines, it was against decades-old federal statutes. None of this made any difference, as there was not adequate enforcement of legal protection for the site. According to the current supervisor of the Spiro site, there still isn't.

When the Pocola Mining Company learned that they would lose their access to the Native artifacts, they dug faster and threw logs and textiles everywhere. University of Oklahoma archaeologist Forrest Clements wrote, regarding the destruction of the main mound, "It was impossible to take a single step in hundreds of square yards around the ruined structure without scuffing up broken pieces of pottery, sections of engraved shell . . . and stone and bone."

Part of the mess left behind by the Pocola diggers was their main tunnel into Craig Mound, the largest of the Spiro mounds. Some local people thought that the tunnel was a danger to local children who climbed around it, so they dynamited the tunnel shut.

After the Pocola looters left, the federal government sponsored a real archaeological dig at Spiro. As part of the New Deal, the Works Progress Administration was eager to finance it. The tools necessary for the project were not expensive, so most of the money could be spent on hiring local people to dig and clean the items they found. Providing employment was, after all, the main purpose of the WPA. Although the Pocola diggers had broken into a chamber that had been sealed for five hundred years, and destroyed many of the perishable items, the WPA found a lot more.

What happened next provides an insight into the attitude of the miners and landowners. This was Oklahoma, which had been Indian Territory, where the chewed-up gristle and bones of nearly every Native tribe in the United States were sent to be confined to reservations. Surely, many people thought, these degraded Natives could not have built so great a civilization. Yet what could these mounds be other than the remnants of a great civilization?

The new white masters did not want to face the fact that Native civilizations might have equaled or exceeded European civilizations. I believe that it was to preserve this mythical image of the degraded Red Man that the Pocola Mining Company destroyed more than they looted, and local people used dynamite on the site. Moreover, the contract with the landowners required the WPA, upon completing their dig, to flatten the landscape so that it would be suitable for farming. This is the reason that, right after the WPA finished, one could no longer see the mounds or even slight undulations of the ground where the mounds had once been. The Spiro Mounds had been completely erased from the landscape, and very nearly from history.

Later, when the land was no longer private property, archaeologists began to reconstruct some of the mounds. You can still see Craig Mound, for it has been partially rebuilt. It is hidden under grass, weeds, and sumac bushes (figure 1.4).

Figure 1.4. Today, the main mound at Spiro Mounds Archaeological Site has been reconstructed. It looks no different from any other little hill in the Oklahoma countryside. *Author photo.*

The destruction of evidence of past Native American cultural achievements continues today, as rock carvings and pictographs in the American Southwest are destroyed by vandals, sometimes by cluelessly selfish individuals, sometimes by white supremacists.[26] While the destruction of Spiro Mounds was the largest example of archaeological destruction, it is far from unique. Throughout the American West, people dig up artifacts to sell, even from graves considered sacred by Native tribes. According to an article in *Newsweek* in 1989, "Until Congress passed the 1979 Archaeological Resources Protection Act (ARPA), unmarked Indian graves enjoyed roughly the status of garbage dumps."[27]

This is why there is almost nothing left of the great Mississippian civilization of North America. The Natives who lived where the Mississippians once ruled forgot about it, and then the whites destroyed most of the remaining archaeological evidence. They paved over the network of trails. Much about the history of Cahokia is unknown; in Spiro, it is unknowable.

It was this continent of clean cities connected by strong trade routes and later by the world's most advanced democracies that was the immediate precursor to European contact.

2

FORGOTTEN FIRES

How Natives Used Fire to Manage the Landscape

Bambi the deer and Thumper the rabbit and all their little animal friends fled from the burning forest. Bambi almost didn't; it took strong words of encouragement from his absent buck father to make him get up and run away from the flames. Careless hunters had started the conflagration by ignoring their campfire.

These are some of the memorable scenes from the 1942 Walt Disney movie, *Bambi*, which among other things portrayed hunters as evil and fires as destructive. But even in this movie, the forest fire was not completely destructive. At the end, Bambi and all his friends (including a bunch of baby Thumpers) gathered together in the charred remains of the trees, around which grew a fresh crop of grass and wildflowers. Despite the best efforts of the Smokey Bear advertisements, most people came to realize that while wildfires destroy, they can also renew natural habitats. I have seen many grasslands and woodlands grow back greener the year after a fire. Wildlife managers, and most nature lovers, recognize fire as a good management tool for natural areas.

WHY FIRE IS ESSENTIAL IN THE NATURAL WORLD

For some natural habitats, fire is not just helpful but essential. Probably every grassland in the world, from North America to Africa to Mongolia, depends on a recurring fire cycle.[1] If one includes tundra in the grassland category, this constitutes over 40 percent of Earth's dry land surface, excluding Greenland and Antarctica. Without fire, humid grasslands would be displaced by forests, and dry grasslands by shrubs. This was especially true of the tallgrass prairie in what became the central United States. Tall—and

34 *Forgotten Landscapes*

I mean tall, sometimes seven to eight feet tall—grasses such as big bluestem and Indian grass dominated this prairie.

Some smaller natural habitats also depend on fire. Some shrubs need to be burned before they can reproduce. This includes the chaparral of California. In the dry summers the chaparral is practically begging to be burned. As I crawled through the chaparral behind Santa Barbara for an undergraduate research project, I could smell the flammable chemicals that diffused from the dry branches and dead leaves. Each of these habitat types contains many species of fire-dependent shrubs. In addition, many pine forest species depend largely or completely on fire for their reproduction. Forests of "closed-cone pines" in California, for example, will simply not open their cones and drop their seeds unless the cones—and thus, the whole forest—have been burned (figure 2.1).

Even within habitats in which fire is not as frequent of an occurrence, some individual plant species require it. In some dry oak woodlands there is a species of wildflower, *Phacelia strictiflora*, that lives for one year. Its seeds remain in the soil, and will germinate only after a fire has passed through. Then it blooms by the thousands (figure 2.2). My research has shown that the seeds require exposure to smoke chemicals in order to germinate.[2]

Different habitats sometimes require different fire frequencies. Fires of about thirty to a hundred years apart maintain the chaparral, and about twenty to a hundred years apart maintain the closed-cone pine forests. Fires more frequent than that will encourage the growth of grasslands.

There are many other habitats that, while not dependent on fire, benefit from an occasional burn. Even part of the Mojave Desert of California burned in 2023. Fire promotes seed production in coniferous forests, even if the trees do not require it. The ponderosa pines of the Black Hills and Rocky Mountains, for example, do not absolutely require a fire for their regeneration. But studies (including mine) of the regeneration patterns of the Black Hills pines suggest an episodic pattern of growth, in which a whole bunch of pine trees begin growing at the same time after a fire.

One of the dominant forms of vegetation in the East was pine forests. In the absence of fire, as noted by Henry David Thoreau during his time at Walden Pond in Massachusetts, deciduous trees such as oaks grew up under the pines (figure 2.3).

Most deciduous forests, without fire, will accumulate dead vegetation which will choke out the seedlings of the dominant tree species. Such a forest would begin to resemble the legendary Mirkwood in the writings of J. R. R. Tolkien. But no forest ever has a chance to become

Figure 2.1. This bishop pine (*Pinus muricata*) forest in the Purísima Hills of coastal California burned in 1995. A nearly pure stand of bishop pine seedlings grew back the following year. *Author photo.*

Figure 2.2. *Phacelia strictiflora* is an annual wildflower of the oak woodland in Oklahoma. It germinates almost exclusively the year after a fire. *Author photo.*

Mirkwood. Everywhere in the world occasionally experiences a long drought, during which dead (and living) foliage can burn. This happens even in tropical rainforests.

When fire occurs, the plant habitat may be destroyed, but the ground that is left behind contains seeds or underground stems that can and do sprout back, usually the very next year. A burned forest I studied in Oklahoma grew back right away as stump sprouts, not only the same species but many of the same genetic individuals as before. For the first few years, the resprouted trees looked like giant bushes. A decade later, one or a few dominant trunks in each clump became the leader shoot of a new tree. Many woody plants, even alder bushes that grow in the middle of shallow rivers, sprout back after fire.

Figure 2.3. In this 1994 photo, hardwood seedlings grow up underneath pines in an unburned forest in Georgia. *Author photo.*

Especially in grasslands, the dark ashes left behind promote new growth. The mineral nutrients such as nitrogen and phosphorus, previously tied up in the dead stems and leaves, are released as fertilizer. The dead stems and leaves no longer shade the soil. This allows new sprouts and seedlings access to bright sunlight and to grow from the ground, and for the ground (now blackened with ashes that absorb the heat of the sun) to warm up faster in the spring.

Immediately after a fire, the most rapidly growing woody species dominate, while slower-growing woody species grow up in their shade and eventually displace them. This happens, for example, in the Black Hills and Rocky Mountains. Aspens grow in very thick stands after a fire. This is because what appear to be separate aspen trees are actually all connected by a vast underground root network. All the trunks are the same size and, after a few decades, begin to grow old and become diseased. Fire clears them away and allows a new crop of vigorous trunks to grow back. Spruce trees start to grow underneath them, and will eventually displace them, unless a fire kills the aboveground aspen trunks, and the whole spruce tree, thus pressing the reset button on the aspen grove.

Given all the benefits of fire to nearly every habitat, fire is a nearly irresistible tool for modern habitat management (figure 2.4). While individual plants and animals die in a controlled burn, the forest or grassland is renewed by the fire.

Most of us realize that, along with global climate change, fire suppression is the reason that wildfires have become so prominent in the news. They are more common and are larger than ever before. A huge wildfire in the Pacific Northwest in 1910 was so destructive that the federal govern-

Figure 2.4. A controlled burn removes the leaf litter and undergrowth from a post oak woodland in southern Oklahoma. *Author photo.*

ment started the National Forest Service in response. One of its missions is to control wildfires on its lands.[3] Recent wildfires have proven ever more destructive, especially the August 2023 fire on Maui and the 2025 fires around Los Angeles. Despite over a century of technological advances in firefighting, by suppressing small fires, we have created a North American landscape that is just waiting to burn.

The very climate of many forests has been altered by twentieth-century fire suppression. For example, in the same oak woodlands to which I referred previously, I have seen red oaks grow up underneath the dominant post oaks. Fires formerly controlled the red oaks, but now they grow up in the shade of the dominant post oaks and eventually replace them.

Also, without fire, the oak woodland becomes clogged with vines such as poison ivy, greenbriers, and grape. The greenbriers, in particular, make a journey through an unburned oak woodland a very unpleasant experience (figure 2.5).[4]

Not all of the forest fires in the past were natural. In literally countless cases, they were deliberately set by Native Americans. They were habitat managers, and pretty good ones, and their main management tool was fire. More on this topic later.

Figure 2.5. In the photo on the left, an oak forest has been cleared out by a controlled burn; in the photo on the right, from a similar oak forest in a different place, the undergrowth has accumulated for a long time. *Author photos.*

40   *Forgotten Landscapes*

AFTER THE ICE AGE

There have been about twenty ice ages during the last two million years in the Northern Hemisphere. Four of them have had a significant impact on the landscape of North America. The most recent ice age ended about fifteen thousand years ago. As I wrote in chapter 1, the American landscape took its present form as the ice melted and retreated northward and up the mountains, where they remain as glaciers. But that was not the only thing that happened. Right about that same time, the first people to ever live on this continent, the ancestors of Native Americans, arrived from Siberia. What happened next was due both to natural and human-induced landscape changes.

Tundra plants and cold-tolerant trees, such as those found today in northern Canada, grew where the ice melted. As temperatures continued to moderate, forests and grasslands developed, in which lived a sparse population of Natives. The tallgrass prairies—which at the time of European contact grew in a north–south band just west of the deciduous forest (e.g., in places that would later become Illinois and Iowa)—established themselves during a couple of millennia in which temperatures were high and rainfall scarce, a period called the *altithermal*. Natural fires maintained these prairies. There was also a moderate level of fires in the forests. These fires maintained oak and hickory woodlands in the drier areas, surrounded by wetter forests of beech and maple.

Next—and this is the part with which most people are unfamiliar—as their populations grew, the Natives started a lot of fires in order to manipulate their habitats. That is, Natives made the prairies, which were already there, bigger,[5] and extended the area of oak-hickory woodlands at the expense of beech-maple forests. Undergrowth was sparse in these forests. It is this phase I primarily describe in this chapter.

Large-scale patterns of temperature and precipitation, not fire (natural or human-caused), determined where the forests, prairies, and deserts grew. The fact that evergreen trees grew to the north of and up the mountains from deciduous forests, and that dry habitats dominated on the leeward sides of mountain ranges, were not the result of Native American fire management. The effects that Native Americans had upon natural habitats occurred *within* the broad patterns established by temperature and precipitation. Whole books have been written about the broad landscape patterns of temperature and precipitation.

Native Fires and Modern Fire Suppression

During the twentieth century, the government tried to control all wildfires. This caused the forest understory to be smothered by thick undergrowth just waiting to burn. The tallgrass prairie was not much affected by fire suppression, since most of it had by this time already vanished. It had been plowed under for cornfields and conquered by urban and suburban human occupation. The only places where you can see tallgrass prairies now are in preserves where the prairie is carefully protected, and burned by habitat managers—for example, the Osage Hills in Oklahoma, where the Nature Conservancy maintains a herd of bison; or along railroad tracks or in pioneer cemeteries, which were established before the rest of the prairie was destroyed.

Since fires make the forest understory drier, they can encourage some trees (such as chestnuts in the Appalachians) over others (such as maples).[6] In chapter 6, I describe how Natives might have enhanced natural chestnut forests by planting chestnuts in them. But chestnuts might also have benefited from the drier understory climate that resulted from fires deliberately set by Natives. One indirect piece of evidence that the deciduous forest of eastern North America was drier and more open in the past than it is today is that the American bison lived nearly through the entire extent of the forest. Such a large animal, and such large herds, could not have lived in the thick forests we see today, but they did live in the more open woodlands maintained by fires, many set by Natives. When Elizabeth was at Mantle Rock in 1838 (see the introduction), most of the trees around her were not the maples that I saw recently, but the post oaks that grow in drier conditions.

MASTERS OF FIRE

Native Americans did not just start fires; they started lots of them, and were masters of its use.

Native Fires Everywhere

Ecological researchers can frequently reconstruct the fire history of a forest, or of patches of trees within a grassland. One way they can do so is to study the rings in the wood of living trees, a science known as *dendrochronology.* Most people know that you can determine the age of a tree by counting

42 *Forgotten Landscapes*

the rings in its trunk, usually from a little cylinder of wood, the extraction of which does no damage to the tree. Each year the tree is alive, it produces a new ring of wood, sometime thick (during good years) and sometimes thin (during poor years). Not only a tree's age but also a record of drought is recorded in thin wood layers.

Tree rings can also provide a record of fires. If a low-intensity ground fire does not kill the tree, it can leave a dark scar in the layer of wood produced the year of the fire. If the tree ring corresponding to the year 1748 has a fire scar, then there was a fire in 1748 at that particular place. The fire history of a forest can be reconstructed in this manner.

This technique, however, cannot distinguish a natural fire from one deliberately set by Native Americans in the past. Circumstantial evidence suggests that Native Americans started fires at certain times of year—for example, early spring or fall, when lightning strikes were less common than in the wet summer (in the middle of the continent), when the fire was less likely to spread. Probably the best evidence that Natives lit fires is eyewitness accounts that literate explorers and pioneers happened to write down once in a while in a journal that might later be preserved in a museum or published.

Only it wasn't just once in a while. It was many times. Whenever any white chronicler saw a Native starting a fire, and wrote it down, the record was available for any scholar to look at and publish. In particular, two scholars amassed a trove of these eyewitness accounts. Starting in the 1950s, Omer Call Stewart[7] began writing about Native American fires. For various reasons I will discuss later, his writings were largely ignored until they were published after his death by Henry T. Lewis and M. Kat Anderson in a book titled *Forgotten Fires*. A forest ecologist named Thomas M. Bonnicksen[8] also amassed a trove of eyewitness accounts and put it in his book, *America's Ancient Forests*. Most of Bonnicksen's examples did not overlap with those assembled by Stewart. I have greatly relied on their compilations in this chapter. So numerous are the examples (173 of them) that I have summarized them into a table in the appendix of this book. Many of the references I mention in the text that follows are in the bibliographies of Stewart's and Bonnicksen's books, which is why I have not referenced them individually.

Stewart may have overestimated the effects of Native fires on the North American landscape.[9] He made it sound as if there would have been no grasslands anywhere if it were not for Native American fires. He seemed to imply also that without these fires, there would be almost no oak woodlands; instead, there would be beech-maple forests. "If there was anything

Forgotten Fires 43

to burn, Indians set fire to it," he wrote. "The whole of the eastern woodlands, from the Mississippi River to the Atlantic, was periodically, perhaps annually, set on fire."

I think most people who have been out in the woods know this is not quite true. But we must consider that the role of Native fires has been greatly underestimated both by scientists and by naturalists. Natives did set a lot of fires; and once a fire got started, what was there to stop it? Their fires might have frequently consumed whole forests. Even if the instances of Native fires were overestimated by early observers, the fact remains that many of these observers actually saw Natives setting fire to the landscape.

Stewart also cites numerous field studies of Native ethnography, published in the early twentieth century, in which Natives told about the extensive use of fire by their ancestors. Stewart conducted some of these studies himself. I have not included them in the table, since they were memories of fires rather than direct observations.

Some of the things you will notice from the table in the appendix are:

- The observations began as soon as Europeans came to North America.
- The observations span four centuries.
- The observations come from all over the continent, not just from the fire-prone West.

The first thing the Natives knew was that fire is uncontrollable. The Natives prepared as best they could before starting a fire, taking into careful account the wind, heat, fuel, and humidity, as would a habitat manager today. Although they did not have the benefit of science, the Natives knew mistakes could happen. Recognizing their own limitations, they did not build permanent buildings (except for stones and bricks in the desert). They built nothing more permanent than a Cherokee longhouse. In some cases this had the added advantage that if a dwelling became too infested with fleas, the inhabitants could simply burn it and build another.

This is in complete contrast with modern Americans. We want taxpayer-funded fire control to keep even natural fires from burning so much as a shed on our property. We assume that our fire-control technology will always work, except when it doesn't. Most of all, we will not accept the idea that there are just some places that are too dangerous to live. Consider, for example, the chaparral of the southern California mountains. We build expensive houses in the foothills behind San Diego and Los Angeles and Santa Barbara, right in the middle of a habitat that

44 Forgotten Landscapes

burns every few decades, and very hotly—and then we get upset when our houses burn. Controlled burns are an especially great liability to government agencies that manage habitats.[10]

The second thing to notice is that the observations began almost as soon as Europeans came to North America. In the sixteenth century, Giovanni da Verrazzano and Álvar Núñez Cabeza de Vaca noted that because of forest fires, which they attributed to the Natives, you could smell the smoke before you could see the land from the ship.

In some cases, Native fires had very considerable effects on the forested landscape, as noted by early European immigrants. P. Lindstrom noted in 1691, regarding Delaware (which is today thickly forested), "There indeed grows a great deal of high grass, which reaches above the knees of a man. . . . There is also no thickly grown forest but trees stand far apart, as if they were planted." Dwight noted in 1822 that the trees that did remain in the large grassy areas of New York showed marks of fire damage.

When George Catlin, the painter who lived among the prairie Natives in the nineteenth century, saw one of their fires, he was impressed with the vast grandeur of the burning landscape:

> These scenes at night become indescribably beautiful, when their flames are seen at many miles distance, creeping over the sides and tops of the bluffs, appearing to be sparkling and brilliant chains of liquid fire . . . hanging suspended in graceful festoons from the skies.

This was a Native habitat management plan, not unlike what wildlife managers do today. These were ground fires, not crown fires; as Adriaen van der Donck noted in 1685, "The green trees do not suffer." This was possible only with frequent, regular burning. Roger Williams noted in the eighteenth century that the Natives considered burning forests to be a benefit.

Later white American observers also admired the way Natives used fire as a habitat management tool. In one of his paintings, Frederic Remington showed Natives setting fire to the prairie (see the cover illustration on this book). In 1870, Major George M. Sternberg[11] watched closely as Natives burned Kansas prairie. He called it the "heroic farming of the American Indian," by which he meant that wild game such as bison were the crop the Natives farmed. Sternberg wrote,

> Indian countries are clean countries. No muddy roads . . . no underbrush or decayed logs and rubbish in their woods, for the annual fires clean up everything, leaving but the greenest trees with thick bark.

The Natives figured out how to use fire as habitat management by themselves; they did not learn it from Europeans. The Europeans who came over on ships were mostly the urban poor with no reason to stay in Europe, or the urban rich who came to seek their fortune. Very few of them were good farmers, or even good hunters. They certainly were not good habitat managers. They had to learn that from the Natives.

The effects of these fires were not always small and local. Even after European diseases had ravaged Native populations, the survivors still used fire to maintain habitats. Botanist William Bartram visited northeastern Florida in 1773.[12] He emerged from a forest of live-oak trees to see "almost unlimited savannas and plains, which were absolutely enchanting; they had been lately burnt by the Indian hunters, and have just now recovered their vernal verdure and gaiety." He saw "herds of deer . . . feeding in the green meadows before us [and] flocks of turkeys walking in the groves around us . . . How cheerful and gay all nature appears." But he also saw some effects of burning that we would recognize as being ecological degradation, mainly soil erosion that "tinge the waters the colour of lye or beer, almost down to the tide near the sea coast." George Emmons noted in 1841 that the smoke could linger a month after a Native fire, enough to obstruct the view of the sky. Fire frequency in the intermountain West was associated with Native agriculture, just as in the East.[13]

It is even possible that fires started by Native Americans may have influenced the evolution of some trees. One study noted that the bark of two species of closed-cone pines in California may have evolved to be thicker in areas that had dense Native American populations than on islands where Natives seldom (if ever) lived.[14] One of the main functions of outer bark, especially in pines, is fire resistance.

WHY NATIVE AMERICANS LOVED TO SET FIRES

I think I could rest my case here, but the whole reason that this chapter is interesting is that the examples show the many ways in which Natives used fire intelligently and skillfully to manage small—or large—areas.

In New Amsterdam, 1685, Adriaen van der Donck saw Natives burning forests along the Hudson River. The Natives were not mindlessly burning the woods. They had well-thought-out reasons, and the Natives told the whites *why* they started fires. "This . . . is done for several reasons," wrote van der Donck:

46   *Forgotten Landscapes*

First to render hunting easier, as the bush and vegetable growth renders the walking difficult for the hunter, and the crackling of the dry substances betrays him and frightens away the game. Secondly, to . . . clear the woods of all dead substances and grass, which grow better the ensuing spring. Third . . . because the game is more easily tracked over the burned part of the woods.

Here are some of the reasons that Native Americans burned their forests and prairies. In nearly every case, the Natives had more than one of these reasons for burning a habitat.

To Promote Soil Fertility

Native Americans, like farmers all over the world for millennia, knew that ashes from crop residues promote the growth of plants the following year. They also knew that grass grows better the year after it burns. Natives set fire to grasslands, or to grassy edges of forests, to promote the growth of grass for grazing and shrubs for browsing, knowing that this would attract deer and other prey, which (since they had no livestock) was their sole source of meat, aside from fish. Sometimes the animals did not wait for the grass to grow. They nibbled at the ashes themselves, which have significant amounts of calcium and potassium salts. (To get sodium and chloride, the animals needed to find natural salt licks or saline water.)

Colonists needed open spaces when they wanted to plant crops, and in 1654, E. Johnson noted that the Natives (in this case, the Wampanoag) saved them the effort: "the Lord having mitigated their labours by the Indians frequent fiering of the woods." As in so many other cases, it should have been the Natives the conquerors thanked, rather than God.

To Clear the Undergrowth

Natives also used fire to clear away undergrowth, to make travel easier. In 1632, Thomas Morton and William Wood in Plymouth Bay wrote:

The Savages are accustomed, to set fire of the country in all places where they come; and to burne it, twize a yeare, vixe at the Spring, and the fall of the leafe. The reason that mooves them to doe so, is because it would other wise be so overgrown with underweedes, that it would be all a copice wood, and the people would not be able in any wise to passe through the Country out of a beaten path.

William Wood made similar observations in 1634. The result was, as John Smith noted in 1624, "so that a man may gallop his horse among these woods anyway except when the creeks or rivers shall hinder." Not to be outdone, Andrew White wrote in 1633 that one could drive a "four-horse chariot" through the forest along the Potomac.

Not only that, but the forests were so open that strawberries grew all over the forest floor, to the delight of Ralph Hamar in 1619 and William Bullock in 1649. Williams enthused in 1650 that "in its season your foot can hardly direct itself where it will not be dyed in the blood of large and delicious strawberries." The strawberries, being very short plants, grew only where the larger undergrowth had been burned away.

As R. W. Wells noted in 1819, regarding the forests of the upper Midwest,

> the Indians travel much during the winter, from one village to another . . . which becomes extremely painful and laborious from the quantity of briers, vines, grass, etc. To remedy these and many other inconveniences, . . . the woods were originally burned so as to cause prairies.

To Gather Wild Foods

According to James Mooney in 1900, Cherokees started fires in autumn leaves in order to more effectively gather chestnuts from the ground. The same was true for acorns. E. Johnson noted in 1654 that fires made acorns more accessible to deer, which the people hunted. But people ate the acorns too. California Natives harvested acorns to grind into a protein-rich meal. As I will explain in chapter 6, they managed and defended their oak woodlands like orchards. They burned the understories of oak woodlands, clearing away other species that would shade out oak seedlings. In addition, the fire would at least partially roast the acorns before grinding.

Another way the California Natives gathered acorns was to let woodpeckers do it for them. The acorn woodpecker (*Melanerpes formicivorus*) not only pecks at and eats acorns, but stores them in tree trunks. (The name *formicivorus* means "ant eater." Acorns are a supplement to insects, which are their main food.) The birds peck holes in dead wood, each hole precisely fitting an acorn, and store them there (figure 2.6). The California Natives would tear down the dead wood and burn it, releasing—and roasting—the acorns.

George Riddle in 1851 reported that tarweeds (*Madia sativa*) produce edible seeds. The problem is, as the name might suggest, that the flower heads have a gummy substance that makes the seeds difficult to harvest.

Figure 2.6. Acorn woodpeckers store acorns in holes that they peck in tree trunks, where they—or Natives—could retrieve them later. *Author photo.*

The California Natives would burn a field with tarweeds. The seed heads remained standing at the tops of the stems. They would harvest the seeds, with the tar burned away and the seeds ready-roasted.

The spring wildflowers of the California coastal grasslands and Sierra foothills, the beauty of which took my breath away when I explored them as a child, were largely an artificial landscape, or at least artificially enhanced. Fires kept out the bushes. According to Father Juan Crespí, back during the California mission days of the eighteenth century, the spring grasslands were

> all one mass of blossom, great quantities of white, yellow, red, purple, and blue ones: many yellow violets or gilly-flowers [phlox] of the sort that are planted in gardens, a great deal of larkspur, poppy, and sage in bloom, and what graced the fields most of all was the sight of all the different sorts of colors together.

The Natives were not trying to simply create a beautiful landscape. They used fire to enhance the landscape into a useful—in this case, an edible—one. Crespí noted, "On this whole march . . . we have not seen a bush." It was the grasses and wildflowers, rather than the bushes, that the Natives primarily wanted. They ate the seeds of California brome, ryegrass, and needlegrass, as well as products from flowers such as chia, spikeweed, redmaid, and brodiaea and mariposa lilies. Cluster-lily *Brodiaea* and *Calochortus* (mariposa lily) have edible corms like their relatives, the onions. According to Moncada, Natives in California would burn the grasslands after harvesting the products they wanted. Today, the grasslands have been largely taken over by European weeds, both grasses and wildflowers. However, many of the original species still bloom in profusion on the spring hillsides.

Shrublands in central coastal California produce few things that the Native inhabitants needed, but they used many products from the grasslands. This part of North America had one of the highest Native population densities on the continent. Perhaps Natives started fires that promoted grasslands at the expense of shrublands. While lightning also starts fires, this region has a very low rate of natural lightning strikes.[15]

Other habitats that required fire had edible and useful products to offer to their Native inhabitants. *Pinus muricata*, one of the closed-cone pines of California described earlier, drops its seeds after fire; the seeds are larger than those of most pines, and very nutritious. Manzanita bushes (genus *Arctostaphylos*) in the chaparral produce edible and much-appreciated berries.

To Promote Hunting

Yet another use to which Natives put fire was in hunting. This was noted as early as 1555 by Cabeza de Vaca. The Natives started fires in circles around places they knew their prey was hiding. As the circle of fire closed in, it drove the prey into small areas where they were easier to kill. This aspect of hunting was especially important before the Natives had horses (which they acquired from Europeans, especially the Spaniards in the Southwest) and could not outrun their prey.

But it wasn't just large prey such as deer that Natives hunted by fire circles. In western North America, Natives used fire to hunt and gather grasshoppers and crickets. The fire circle would not only concentrate the insects, but would scorch their wings and legs, leaving prey that was more edible (perhaps even cooked) and which could not fly or jump away. In some cases, they built fires over wasps' nests, which would drive away the

50   *Forgotten Landscapes*

adult wasps and cook the larvae still in the nest. In California, the Natives would mix the cooked grubs with ground acorns and manzanita berries before eating them.

One hunting advantage that resulted from Native fires was that it cleared away dry understory plants which, if trodden upon, would frighten the game away. R. W. Wells wrote in 1819, regarding upper Midwest forests, "If the woods be not burned as usual, the hunter finds it impossible to kill the game, which, alarmed at the great noise made in walking through the dry grass and leaves, flee in all directions at his approach."

George Davidson reported in 1879 that Natives would burn aquatic vegetation around lakes, causing it to grow more rapidly the next year. This would attract ducks. The Natives had to burn the vegetation at the right time. They made sure they did not burn duck habitat after the beginning of the nesting season. He also noted that fires in open areas of a forest promoted the growth of berries, a valuable source of food for the Natives. The berries also attracted bears, which the Natives would hunt.

The Natives thought ahead not only to the next spring but to later years. Prairie tribes moved their camps to follow the bison. The bison went to places with fresh grass—that is, to places the Natives had burned during a previous year. Some tribes burned aspen groves, causing them to grow back and attract beaver.

To Obtain Non-Food Products

Natives had numerous uses for conifer sap—for instance, to help attach arrowheads and spear tips to shafts. They would burn certain kinds of coniferous forests and gather the sap that boiled out of the trunks.

California Natives were famous for making baskets that could hold water, even boiling water from hot rocks immersed into it. They made these baskets out of the stems of shrubs such as dogwood and the leaves of aquatic sedges and cattails. Old stands of shrubs and sedges produced short and inferior material for basketry. If the Natives burned the shrubs and sedges, the next year's growth would be rapid, straight, and long. The burned shrubs would also grow long, straight stems that they used for arrow shafts.[16]

For Pasture

Even before migrating to Indian Territory, what is now Oklahoma, the Cherokees had discontinued many of their traditional ways. At first glance,

Cherokee farms in what is now Oklahoma looked the same as white farms. But they continued some of their previous habits of thought. Perhaps the most important concept was that nobody owned land. As I explained in the previous chapter, land belonged to the whole tribe. Individuals and families could own the things on the land, such as houses, horses, crops, and tools, but the land itself belonged to the whole tribe. This practice continued until the federal government forced the tribe to divide up the land and allot it for individuals to own. All the farmers, including my Cherokee grandfather Edd Hicks, would get together and help burn one another's pastures, a task too big and risky for any individual farmer.

Escaped Fires

When a Native village moved to a new location, they took fire with them in the form of "slow matches" made of dry plant fibers burning at one end. But they seldom completely extinguished the fires in their old habitation. Instead, they would bank them with dirt, which allowed slow combustion to continue. Sometimes the fire would escape and burn nearby habitats. This was not, however, often considered a problem. If the village planned to come back to the same location, they might be glad to find the fire still smoldering, especially considering that the slow match, despite caution, might get extinguished by heavy rain or by the fire-bearer falling in a river.

For Communication

Nearly every tribe used smoke signals for rapid, distant communication. However antiquated this seems today, it was the equal of European and American communication at the time. In Europe, information was communicated from one mountaintop fort to another by torches in windows, essentially the nocturnal version of smoke signals. It was the technology of "One if by land, two if by sea" used by Paul Revere in the Revolutionary War.

For Safety

Another reason that some Natives burned forests and grasslands was to promote visibility. Safety demanded it. Enemy soldiers, white or Native, could hide in thick undergrowth. George Catlin noted in 1832 that the grass in the tallgrass prairie could get so tall that a man on horseback couldn't see

52   Forgotten Landscapes

over it even if the rider stood in the stirrups. Not only could an enemy hide in the grass, but one could not see where he was going. It was not just humans. Grizzly bears could also hide in California chaparral bushes.

According to several writers in the table in the appendix, from the Midwest to Oregon, Natives would deliberately burn the grass to prevent white soldiers from having access to forage for their horses. This was at a time when these Natives did not yet have horses.

Often the Natives would start fires in order to protect themselves from future fires. Natural treefalls could self-combust and present a fire danger in the hot, dry summer. Natives would burn the treefalls in the spring, reducing the danger of summer wildfire. In at least one Southwestern pueblo, frequent Native use of fire reduced the combustibility that might otherwise have threatened the pueblo.[17]

POLLEN IN THE MUD: A WINDOW INTO HISTORY

Scholars know that fire was, and is, essential in maintaining the great prairie at the center of North America. But it is quite another thing to say that, without fire, there would be no prairie at all. It is likely that Native American fires at least expanded the area covered by the tallgrass prairie. This is especially true about what early-twentieth-century plant ecologists called the "prairie peninsula"—that is, the grassland that extended eastward into Illinois and Wisconsin, into climate zones otherwise occupied by forests. The prairie peninsula, at least, would undoubtedly have been forest were it not for Native fires.

The only way, and then only roughly, to know what kinds of plants covered the landscape prior to direct observational records is to study the pollen grains that accumulate in mud. Each kind of plant (often, each genus or family) produces recognizably different shapes of pollen. Since pollen has to keep reproductive cells alive as it blows or is carried through the air, it has a strong outer covering. Even long after the pollen grain itself is dead, this covering can persist in mud at the bottom of lakes. Scientists are thus able to reconstruct a vegetation history of the landscape surrounding the lake by digging down into the mud.

James King, a pollen expert who worked for the Illinois State Museum, reconstructed the pollen history of Chatsworth Bog in Livingston County, Illinois, in the prairie peninsula. After the most recent ice age, most of the pollen was from trees, such as spruce, that characterize the northern forests now found in Canada. Later, oak pollen accumulated, an indicator of deciduous forest. Every habitat has grasses, but only grasslands produce enough grass pollen to have a noticeable effect on the pollen profiles. King identifies 7,900 years ago as the beginning of the prairie peninsula vegetation, with its drier conditions. By about 5,000 years ago, wetter conditions might have brought an end to the prairie peninsula, but prairies persisted, probably because Natives were burning the grasslands, thus suppressing the trees. These results suggest that Native fires did not create the prairies but rather maintained them.[18]

What Happened When the Natives Stopped Burning

It would be helpful, one might suggest, if there was an experimental test: Remove the source of the fires (that is, the Natives) and see whether the prairie persists. In this experimental design, if you remove what you think is the cause, then what you believe to be the effect should also disappear.[19] This cruel and inhumane experiment has been carried out. Natives were driven from prairie lands in Wisconsin and Illinois, and the white settlers who replaced them did not continue the practice of regular burning. Within a century, forests grew where prairies had formerly dominated. In 1829, 90 percent of southwestern Wisconsin was prairie, but only 30 percent by 1854. The first settlers who came to Wisconsin, after the expulsion of the Natives, had a hard time finding firewood.[20]

During the relatively short time in the nineteenth century when forest replaced prairie in the prairie peninsula, the climate and soil changed very little. The only drastic and fundamental change was the departure of the Natives and the arrival of the whites. These new forests grew on what soil scientists recognize as prairie soil.

Once European diseases had devastated Native populations, and they stopped burning their forests and prairies, the effects were immediately obvious. The forest understory grew back, and quickly. In 1629, William Wood said that there was lots of undergrowth in places where Natives had died of European diseases and no longer maintained their forests.

This also happened after Natives were ejected from their ancestral lands and sent to reservations. A British geologist, George W. Featherstonhaugh, visited the Cherokees in 1837, just before the Trail of Tears, and noted that the forests were clear of undergrowth. When he returned in 1844, after the Cherokees had been ejected, he found that the forest undergrowth had returned.[21]

54   *Forgotten Landscapes*

WHY THE IMPORTANCE OF NATIVE FIRES
WENT UNRECOGNIZED FOR SO LONG

My first impression, when looking at the evidence collected by Omer Stewart, was to consider that it led to the inescapable conclusion that Natives skillfully managed the North American landscape using fire as their tool. The matter should have been settled in the 1950s, when Stewart presented his results. Instead, it took until a half-century later for scholars like Thomas Bonnicksen, Henry Lewis, and M. Kat Anderson to get this concept noticed among academics, and even longer for popularizers such as myself to write about it.

Omer Stewart was an anthropologist. At the time he was presenting his conclusions, anthropology was dominated by the concept that human societies progressed through a series of stages until reaching their peak in modern white civilizations. The first stage was when savage barbarians lived in and from the land, without significantly altering it—certainly without deliberately managing it. Only with the origin of agriculture, according to this view, did humans begin to manipulate their environment. Led by prominent scholars such as Julian Steward, anthropologists assumed that Natives were savages and could not possibly have managed their habitats.

This viewpoint was immune to evidence, not only of Native fires, but also of Native agriculture (chapter 4). A representative of the California Division of Forestry noted, with regard to Omer Stewart's as-yet-unpublished ideas, "It would be difficult to find a reason why the Indians . . . should care one way or the other if the forest burned. It is quite something else to contend that the Indians used fire systematically to 'improve' the forest. Improve it for what purpose?" Julian Steward and Omer Stewart often faced off in court cases in which the federal government had to decide what land use rights to extend to Native tribes. Steward testified that Indians were not fully using the land and did not deserve to keep it, while Stewart defended (always successfully) Native land claims.

Stewart and Steward even worked on one of the same New Deal government projects, called the Cultural Element Distribution project. Anthropologists gathered memories from Natives about whether and how often their tribes set fire to the landscape, and why. In his unpublished book, Stewart used Steward's own words against him. Steward's own 1948 report showed that Natives repeatedly set fires to parts of the landscape and knew why they were doing it.

Academic publishers declined Omer Stewart's manuscript. There might have been some good reasons; for example, his writing was appar-

Forgotten Fires   55

ently jumbled and very long. The academic rejections seemed pretty brutal, however, especially as one of the publishers lost the manuscript for two decades before rejecting it. (One of their reasons was that the references were all over two decades old.) Stewart needed the help of colleagues, but didn't get it until Henry Lewis and M. Kat Anderson cleaned up his manuscript and published it after his death.[22]

The period of benign neglect continues. Lewis and Anderson may have rescued Stewart's book, but it's hard to find today. Though many scholars in environmental biology have done their own research on Native American fires, this topic is a backwater in ecology literature, and not very noticeable in anthropology literature.

Stewart's writings opened my eyes to some new possibilities. One of these has to do with the timber- or tree line. The forests that grow farthest north are the boreal forests, and those that grow the furthest up a mountain are called subalpine forests. At the timberline, the trees stop growing and tundra takes their place. The tundra looks like a grassland, but consists mostly of sedges and low bushes. The general story—at least the way I always taught it and wrote about it—was that, north of or above the timberline, the winters were too cold and the soil too shallow for trees to grow upright. Below the soil surface, whether a few inches or a few feet, the soil was frozen (permafrost) and, as far as roots were concerned, might as well be rock.[23]

The evidence for this is that the trees right at the timberline were short and often damaged by winter winds. They might have branches growing on just one side of the trunk, the leeward side, as winter winds would have picked up little shards of ice and destroyed buds on the windward side. Down near the ground, winter snow protected the growing tips of the low shrubs and shrublike trees. It looked pretty obvious to me and to most other students of nature that the tundra was maintained by the forces of winter.

The problem is, some of the evidence didn't fit. Stewart, with his eyes peeled for evidence of fire, noticed that some tundra areas contained pieces of charred wood. While at soil level, the tundra is always wet, on a windy summer day (when temperatures can get surprisingly warm), fire can spread and burn any vegetation that is not near the ground. This would keep trees out of the tundra as effectively as winter winds. The truth is probably that both summer fires and winter winds keep trees out of the tundra. It would seem unlikely that Natives were setting fire to the tundra in summer—unless they were encouraging the growth of small plants to feed caribou, one of their preferred prey.

56   *Forgotten Landscapes*

Even the forests of the Pacific Northwest have been strongly influenced by fires. Some of the tall coniferous trees live for many centuries, especially the coast redwoods (*Sequoia sempervirens*). These trees grow only in a zone of coastal fog, which slows down the loss of water from their foliage. This is the only way that water can reach the top of the trees, which are the tallest in the world. The most abundant trees in Pacific Northwest forests are Douglas firs, which live for about two to three centuries. Many ecologists believe that these forests burn every few centuries, and this allows Douglas firs to begin growing. Underneath them, other conifers, such as hemlocks (as well as a few hardwoods, like bigleaf maples), begin to grow. Were it not for fires, the hemlocks and other shade-tolerant conifers might eventually dominate the forests.

The Northwest has a lot of lightning, but it is not clear that lightning could have started enough fires to maintain the Douglas fir as an important tree prior to white contact. Of course, the Natives did not set fire to the forests in order to get Douglas firs to grow. They did use timber, though they preferred the rot-resistant cedars of the genus *Thuja*. They did start fires in openings to promote the growth of berries, and once in a while these fires must have gotten out of control. The Natives might have used fire as a management tool for berries, but the fires ended up having a landscape-level effect on the forests.

VISIONS OF HEAVEN: FIRE-MANAGED LANDSCAPES

Coastal Grasslands in California

I experienced some of the fire-enhanced beauty of the California foothills for myself. In 1981, I drove along the Pacific Coast Highway in California. I stopped near a big bluff that swept below the highway, and ended in cliffs against which large waves crashed. It was so beautiful that I literally ran and slid down the bluff through a wonderland of wildflowers and grasses, a mixture of colors not unlike those that uplifted Juan Crespí centuries earlier: scarlet Indian paintbrushes, yellow lotus, and large, pale-purple seaside daisies. I was very happy that coastal California could be so beautiful.

In 1988, I drove along that same highway and recognized the very same bluff I had traversed seven years earlier. Only this time, I had to crawl through thick brush, especially the dense coyote brush (*Baccharis pilularis*). On my way down to the cliffs at the bottom, I did not see any of the wildflowers that had so delighted me on my first visit. On the way back up, I found myself assaulted by a thick growth of Pacific poison oak (*Toxicodendron diversilobum*), a dermal sample of which I took back with me to my home on the East Coast.

The difference between 1981 and 1988 was that my first visit had been a year or so after a fire had burned away all the shrubs and poison oak, which would have choked out the grasses and wildflowers. The bushes grew back during the ensuing seven years. Had I been a member of a Native California tribe, I would have found the 1981 landscape not only beautiful but bountiful, and the 1988 landscape hostile. And my tribe would have set fires to make sure that there was as little of the unpleasant shrubland as possible.

Giant Sequoia Forests

Each time I visit Sequoia National Park in California, it is like a vision of Heaven. This is a national park that has been set aside not only because it has the loftiest peaks of the Sierra Nevada, but because it is home to the giant sequoia trees (*Sequoiadendron giganteum*). Though most of the trees are individually named after Civil War heroes, including Confederate ones, the species itself is named after the Cherokee scholar Sequoyah, who almost single-handedly invented the Cherokee writing system.

As you drive up the mountain toward the sequoia forest, you circle around one bend in the road and go from a hot south-facing slope to a cool north-facing one. You are instantly blanketed by shade and scent from the incense cedars and firs. In a few isolated locations in this coniferous forest you can find groves of giant sequoia trees.

To stand next to one of these giant sequoia trees is a humbling experience. The trunks have bases as big as houses, the diameter undiminished for a couple of hundred feet vertically. They cast a cool though not gloomy shade over the layers of shed bark and dead cinnamon-colored needles. You can see plenty of sky between the tree canopies (figure 2.7). The understory is open, and you can walk anywhere without difficulty (though you should remain on the designated Park Service trails). There are plenty of understory plants, from small trees such as California dogwood and Kellogg oak, to bushes such as chinkapin and gooseberry, to wildflowers such as lupine and, in June, the astonishingly beautiful *Linanthus montanus* phlox. The air is delicious.

Though I did not realize it when I first visited, this forest resulted from a cycle of fire. Fire allowed the understory to grow, but not too

Figure 2.7. Giant sequoias (*Sequoiadendron giganteum*) are the largest trees in the world (left). The large trees cast deep shade but are sometimes widely spaced (right). *Author photos.*

much. Sequoia seedlings do not germinate well in the shade. Fires open up sunny spots in which the seeds can germinate. On the trail around Crescent Meadow, you can see several fire-charred sequoia trees (figure 2.8). Many of the larger trees continue their healthy growth because the thick bark protects the growth layer in the trunk from damage from all but the biggest fires. Some of the trees were dead, some just barely alive. It was in the vicinity of these fire-charred trees that I saw young sequoias growing. This was the kind of sequoia forest that John Muir saw at the end of the nineteenth century, and that Native Americans would have burned before European and American contact.

Increasingly, it is not the kind of forest you see today in other parts of the park or in adjacent areas of national forest. White fir (*Abies concolor*) has always been a component of the forest, but in the absence of fire, it produces dense stands of small trunks. When white fir trees die, after a life much shorter than that of a sequoia, they contribute to a thicket of flammable wood.[24] When a fire inevitably occurs, it becomes a "big one," with lots of fuel. This happened in 2021, when lightning started some fires in which some of the biggest sequoias were on the national news because park personnel waged a desperate battle to save them by placing heat-reflecting

Forgotten Fires 59

Figure 2.8. The sequoia forests results from a cycle of fires. Young pointed sequoia trees grow up under a mature tree, while a charred sequoia stump stands nearby. *Author photo.*

blankets at their bases. Severe fires are now killing giant sequoias that, in the past, survived the more frequent and more moderate fires.[25]

WORLD IMPACT OF NATIVE AMERICAN FIRES

The Natives set so many fires that they made a significant contribution to global atmospheric carbon. When diseases killed many of the Natives, the forest undergrowth grew back and the resulting photosynthesis removed a lot of carbon from the air. According to one study, this decrease in heat-trapping gases was enough to cause the Little Ice Age, a period of cold weather in the Northern Hemisphere.[26] It is one thing for Featherstonhaugh to observe that the departure of Cherokees, after the Trail of Tears, resulted in the growth of underbrush, and quite another for the leader of this study to say that the decline in Native fires changed Earth's climate. The study estimated that 56 million Natives died between 1492 and 1600 due to disease. This caused a rapid reduction in Native American fires. The resulting 55.8 million hectares (138 million acres) of reforestation would have removed 7.4 billion metric tons of carbon dioxide from the

60   *Forgotten Landscapes*

atmosphere, enough to cause worldwide climatic cooling. This conclusion is not universally accepted among scientists.

THE FORGOTTEN LEGACY OF NATIVE FIRES

Nearly all natural ecosystems have evolved with fire. Fire suppression, especially to the technologically sophisticated and well-funded extent that our modern culture demands, is unnatural. All plants have adaptations to fire. Fire resistance is one of the reasons that bark evolved in the first place on woody plants hundreds of millions of years ago. All over the continent, Native American fires enhanced the natural role of fire and allowed the Natives to share its benefits.[27]

Joaquin Miller, "Poet of the Sierras," summarized the Native use of fire: "Fire was always the servant, never the master." Not only did it enhance the material needs of the tribes, but it also produced landscapes that we always assumed were totally natural. We admire the open spaces of Yosemite Valley, but according to Galen Clark, it originally resulted from burning by Miwok Natives.

The conclusion is this: The Natives just wanted to maintain their way of life, and fire was one of the tools they used. But there were landscape-level effects, like making the prairies bigger and the forests drier. This was perhaps the major way in which Natives transformed the North American landscape.

3

MASTERS OF THE HUNT

How Native Hunting and Fishing
Controlled Animal Populations

Not only did Native American fires transform the landscape, creating clear forests and open prairies, but their hunting activities also helped to keep some wild animal populations in check because they practiced what we would today call wildlife management. Unlike modern wildlife scientists, Native Americans were not trying to manage prey populations. They were just looking for something to eat. Or were they? Native hunting and fishing practices had some characteristics that suggest deliberate and conscious attempts to protect prey populations from overexploitation.

ANIMALS HUNTED BY NATIVE AMERICANS

In this chapter we will examine examples of animal populations that Native Americans hunted. In some cases, the animal populations exploded when Native populations declined, principally from European diseases, as I will discuss in chapter 7. And in other cases, they did not.

Astonishing Populations of Passenger Pigeons

Most people who have read about American history have been astonished by the accounts of the passenger pigeon in the Eastern deciduous forest during the eighteenth and nineteenth centuries. Were the stories not so numerous or widespread, they would be considered unbelievable. Thick clouds of passenger pigeons, dark enough to obscure the sky, stretched for many miles. At any one time, according to some scientific estimates, there were four to seven billion passenger pigeons in North America, and they made up as much as 40 percent of the birds in North America.[1]

62   *Forgotten Landscapes*

In 1623, Gabriel Sagard-Théodat recorded "an infinite number of turtle-doves." Cotton Mather in 1714 witnessed a flock a mile wide that took several hours to pass overhead. In 1813 John James Audubon, the famous birdwatcher and painter, saw a huge flock of passenger pigeons. "The air was literally filled with Pigeons; the light of noon-day was obscured as by an eclipse; the dung fell in spots, not unlike melting flakes of snow, and the continued buzz of wings had a tendency to lull my senses to repose," he wrote. Audubon saw them during an entire day of fifty-five miles of travel. "The Pigeons were still passing in undiminished numbers and continued to do so for three days in succession," he wrote.[2]

Audubon was not the only nineteenth-century observer to describe pigeons migrating overhead from morning until night for several days. A single 850-square-mile nesting area in Wisconsin was estimated to have 136 million birds. In 1866, one flock in southern Canada was almost a mile wide and three hundred miles long, took fourteen hours to fly over, and consisted of more than three and a half billion birds.[3] Regardless of the accuracy of these pigeon population sizes, these are big numbers we are talking about.

Some scientists have speculated that when there were billions of pigeons, they may have had an impact on the kinds of trees that dominated the oak forests. Pigeons would eat acorns of both red and white oaks. White oak acorns, however, fell in the autumn and germinated, overwintering as dormant seedlings. In the spring these seedlings were less visible, making them less of a food source for pigeons. Red oaks, however, dropped their acorns in the spring, where newly migrating flocks of pigeons could easily find them. Over the years, this may have made white oaks more common.[4]

And this, many have assumed, was the way it had always been, at least since the forests grew back after the most recent ice age. This is what I believed, too.

But there was something not quite right about this picture. These huge populations could only happen if there was a nearly unlimited amount of food for the pigeons. During the summer, the pigeons ate mostly fruits—not only wild fruits that humans find delicious, such as wild plums and blackberries, but also fruits we usually avoid, such as dogwood berries. The rest of the year, they ate mostly nuts, such as beech nuts and acorns, as well as chestnuts from extensive forests that have since vanished because of a disease called chestnut blight. Pigeons also ate small invertebrates such as earthworms, caterpillars, and snails. They had a variable enough diet that they did not starve from lack of food even during times of peak population size.[5] They also ate cultivated grains, mostly buckwheat, but they have not

Masters of the Hunt   63

been recorded as a major crop pest. This made me wonder, as you have probably already guessed: How could there have been enough food for billions of pigeons over thousands of years?

In 1850 there were as many pigeons as ever, but by 1900 it was nearly impossible to find them in the wild. The last passenger pigeon, Martha, died in the Cincinnati Zoo in 1914. Despite the massive number of pigeons killed by nineteenth-century professional hunters to supply a demand for pigeon meat in the East, a delicacy known as squab, it seems impossible that hunting could have reduced their numbers so drastically and so quickly.

But there were other issues that contributed to the crash of pigeon populations, followed by extinction. One such factor was that passenger pigeons lived and bred in forests, and depended principally on forest foods. In the eighteenth and nineteenth centuries, American forests were extensively destroyed for farming, cities, and industry. With less forested area, the decline in pigeon populations was inevitable—but not the quick decline observed between 1850 and 1900.

In addition, researchers have presented genetic evidence that passenger pigeon populations fluctuated drastically in size during the course of many centuries.[6] A natural population crash exacerbated by hunting and forest habitat loss may have been the cause of extinction. That is, this was an example of multiple causation, the rule rather than the exception in the natural world.

This leaves another question unanswered. How could such a large pigeon population have developed in the first place? Although white observers, with their journals, were rare in North America before 1600, there seems to be no evidence of prodigious pigeon populations between about 900 BCE and European contact. Was there something that kept pigeon populations relatively low before European contact?

The answer to this question might be the same answer we encountered in the previous chapter: Native Americans kept pigeon populations low, just as Native fires kept forests and prairies clear and open. How did they do this?

One researcher says the answer is competition. Natives ate so many wild nuts that the pigeons could not find enough food for huge populations.[7] But another explanation, hinted at by many researchers, is that Native Americans kept pigeon populations down by hunting them. Certainly pigeons were a common food source for Native Americans. Pigeon bones, though not billions of them, were abundant at Native archaeological sites.[8]

So how did Native Americans keep billions of pigeons under control? Quite simply, they didn't have to. Some basic arithmetic will demonstrate

64   *Forgotten Landscapes*

this. For purposes of calculation, let us begin with a pigeon population of 300,000, which can grow by 10 percent each year unless there is some natural disaster such as a drought. If Natives hunted about 10 percent of the pigeons each year, the population would remain at 300,000 indefinitely. But if the pigeon population actually grew at 10 percent a year, it would take only eighty-seven generations for there to be over a billion of them. You are welcome to do the calculations by hand, as I did, not being a competent mathematician. I might have gotten the numbers wrong, as Darwin, also not a great mathematician, did in his calculation of elephant population growth in *The Origin of Species*. In an equilibrium situation, before a population explosion begins, Native hunting could easily have kept pigeon populations low. After a population explosion begins, almost no amount of hunting—even blasting the trees where the pigeons roosted with ammunition, as white hunters did in the nineteenth century—can control it.

Nearly every adult Native male was a hunter, but it took many years of practice before a Native man could hunt big game, such as deer and bison, very dangerous animals in addition to being difficult to subdue. Pigeons, on the other hand, were easy prey. Pigeons were an important Native American food source, second only to the turkey as a source of meat. Pigeons store a lot of fat, and Natives would save the fat (which could become rancid but would not rot) and use it as butter.

Native hunters would kill pigeons in their large colonies using long poles at night. When the pigeons flew low, Natives could kill them with sticks or stones. Pigeons could be killed with a single small arrow shot—or a projectile thrown—by a little boy. While their fathers were off hunting a relatively small number of big-game animals, the sons would hunt a large number of small animals, and could feel the satisfaction of contributing to the food supplies of their family and the village. The fact that the pigeons flocked together meant the boys did not need to search far and wide for them. While nesting sites varied from year to year based on the irregular patterns of fruit and nut production in the forests, the whole tribe could keep track of where the pigeons were nesting or roosting at any given time.

Native Americans were clearly not the only predator of passenger pigeons. Many predators sought pigeon meat, particularly of nestlings. Minks, weasels, martens, and raccoons could climb on the branches that held the nests. Larger predators, unable to climb far enough into a tree canopy, could eat nestlings that fell on the ground. On the other hand, predatory birds seldom had opportunity to catch enough pigeons to have a significant effect on their populations. The pigeons, which could fly at sixty miles per hour, were light and sleek. Their social behavior was sophisticated enough that an entire flock on the wing could swirl and escape pursuit, an almost mesmerizing behavior observed by Audubon and many others.

But of all the animals that preyed upon the passenger pigeons, there was only one that suffered a severe population decline, starting before 1600. That was the Native Americans, mainly as the result of European diseases. The other predatory animals died in large numbers at the hands of the whites, but this process did not intensify until well after the pigeon population explosion had begun. In fact, most of the predators that were killed by white hunters died in the early twentieth century, most notably wolves, as a result of government-sponsored eradication campaigns. Until that time, the deaths of Native Americans may have left more food for other predators.

My conclusion is that the breathtaking population explosion of passenger pigeons coincided in time with, and was partly caused by, the die-off of Native Americans, who had previously been one of the major predators upon the pigeons.

What about the Deer?

Pigeons were not the only prey of the Native Americans. Another major source of meat was venison. Native Americans and deer hunting are closely allied in popular imagination. One might therefore expect that when Native populations crashed, there would have been a population explosion of deer like that of the passenger pigeons. If this occurred at all, it was not significant enough for anyone to notice. If I invoke Native hunting as the force that controlled passenger pigeon populations before European contact, should I not also have to do so for deer?

It is likely that pre-contact deer populations, like those of the passenger pigeon, were kept in some kind of equilibrium before the Native die-off. What happened next, however, is a little more complicated. As in earlier chapters, I will use the Cherokee tribe as an example.

Deer hunting was practically a rite of passage for Cherokee males, as for males of many tribes. Epidemics, primarily smallpox, reduced the Cherokee population by about half, especially in the plague of 1738. This would inevitably have led to a massive decline in the number of hunters.

But another factor was at work about this same time. Before European contact, deer were a carefully hunted source of meat and skins for the Cherokees. By about 1700, there was a sizable market for animal skins, including deerskins, in Europe. Profit was enough of an enticement that many European—and later, American—hunters and trappers set out in search of animal skins they could sell.

66   *Forgotten Landscapes*

Cherokee hunters already knew how to hunt deer very skillfully. White traders found it profitable to buy deerskins from Cherokee hunters. White traders after 1711 happily sold guns to Cherokee hunters to enhance the number of deerskins they could get. (They did not sell rifles to the Natives. Rifles have spiral grooves in the gun barrel that improve the aim of the bullet and would allow Cherokee warriors to more accurately shoot white soldiers.) Cherokee hunters killed far more deer than they needed for food and for their own clothing. Thus, about the same time the number of Cherokee hunters plummeted, the number of deer killed per hunter increased. These two processes may have partially canceled one another out. Deer went from being a respected prey animal to being a commodity for the Cherokee tribe.

Moreover, humans were not the only, or even the major, predator on deer. Wolves, which were widespread in North America, were presumably only too happy to kill whatever deer the humans did not, as they always had. Wolves would also have prevented the deer from having a pigeon-like population explosion after European contact. Humans may not have been as important of a population control factor for deer as they were for pigeons.

The importance of wolves in controlling deer populations is shown by what happened when white settlers tried to eradicate wolves. European and white American settlers wanted to do whatever they could to prevent predators from killing their livestock. Before 1900, all of these efforts, which focused on killing predators, were private ventures. Groups of ranchers would offer bounties for every wild predator that someone killed. After 1900, especially when the demand for red meat increased during World War I, the Bureau of Biological Survey (predecessor of the Fish and Wildlife Service) of the federal Department of Agriculture began to not only offer bounties for killing predators, but also hired employees to kill predators on federal lands, such as national forests. They also killed wolves in order to allow the deer populations to increase, for the benefit of hunters.[9]

The government accomplished this task only too well. By about 1930, observers were becoming alarmed that government predator control efforts were going too far. One young government employee, Aldo Leopold, enthusiastically shot wolves until one day, the story goes, he watched a wolf that he had shot die:

> We reached the old wolf in time to watch a fierce green fire dying in her eyes. I realized then, and have known ever since, that there was something new to me in those eyes—something known only to her and to the mountain. I was young then, and full of trigger-itch; I thought

that because fewer wolves meant more deer, that no wolves would mean hunters' paradise. But after seeing the green fire die, I sensed that neither the wolf nor the mountain agreed with such a view.

Leopold later became the principal founder of the science of wildlife management, as I will describe later.[10]

One of the results of the attempted eradication of wolves was a population explosion of deer. In fact, there were so many deer on the Kaibab Plateau north of the Grand Canyon that, starving, they chewed on aspen sprouts. As a result, there was practically no new aspen growth during those years. There was, in fact, a population explosion of deer, but it was only after a massive wolf eradication effort.[11]

As is the case with every population explosion, including the pigeons discussed earlier, the population crashes as it overshoots its food supply and also becomes the victim of diseases. The deer population explosion was followed by a deer population bust. At this time, a massive aspen sprout began. When I visited the Kaibab Plateau in 1981, everywhere I looked there were masses of aspens.

The Case of the American Bison

By the eighteenth century, hundreds of millions of bison grazed the prairies and plains of North America. There might have been a large number of them in the forests, as well, back when Native fires cleared away the undergrowth. But by the 1700s, bison were rare in Eastern forests, although Robert Beverley wrote in 1722 that there were bison in Virginia in the open fields. In most of the Eastern forests, after the discontinuation of Native fires, the undergrowth was just too dense for the bison to easily graze or gallop. But such a large population of bison in the prairies during the 1700s invites us to investigate whether this was a natural and sustainable population, or if it was a population explosion that resulted from the deaths of Native hunters who had previously controlled their populations.

It is clear that hunting bison required the highest possible skills for Natives who did not at first have guns. The image of the Native horseman chasing down a galloping bison and shooting with arrows or spearing it is one to inspire awe. It required superlative horsemanship, especially since the bison were very large and very dangerous animals. Once one or more bison were killed, every part of the animal would be used: the meat for food, the hide for leather clothing and tents, the bones for tools. The entire culture and religion of the Plains tribes centered on the bison.

68 *Forgotten Landscapes*

Hunting bison was even more difficult and dangerous before the Natives had horses, which they got from the Spaniards in the Southwest before 1600. Working in closely coordinated groups, the horseless hunters would sneak up on an isolated bison, keeping downwind, and often covered with the hides of recently killed bison. Only at the last minute would the hunters emerge and attack the prey. Sometimes the Native hunters would guide the bison toward cliffs, down which the bison would tumble to their deaths. Although the hunters would usually hunt only what they needed and use all of what they got, it was not always possible when forcing a herd over a cliff to kill just the number of animals they needed at that time. But when horses became available, the Natives were able to kill a lot more bison than they could before.

Although the Plains Native populations were devastated by European diseases, there still seemed to be plenty of bison hunters in the eighteenth and nineteenth centuries. This is partly because some of the tribes that we most closely associate with bison hunting, such as the Dakota, actually lived in the Eastern forests until not long before they began to hunt bison. Pushed westward by other tribes, who had themselves been pushed westward by white immigrants, the Dakota and other tribes arrived in the prairies about the same time that horses became available.

The only way to know for sure whether or not the huge populations of bison had been present for many centuries or even millennia, or whether they represented an explosion resulting from the deaths of Native hunters, is to have some eyewitness accounts of bison populations before European contact. Such eyewitness accounts are rare, but they do exist. In particular, Francisco de Coronado, while traveling in the North American grasslands, reported seeing herds too big for him to estimate their sizes. More than this, he saw these herds day after day as his group traveled. It is unlikely that he was seeing the same herd over and over. He reported his observations to the king of Spain in 1541. This was only about a half-century after Columbus. The bison populations, I conclude, had been large for millennia and did not result from a spurt of overpopulation.

We can never know if the destruction of Plains Native populations might have produced a bison population explosion, because at about the same time, white hunters were killing as many bison as they could and changing the landscape so that it was no longer open to bison migration. In fact, the destruction of the bison was part of a deliberate process by which the United States government tried to undermine the economy and independence of Plains tribes. Sometimes trains transported amateur white hunters across the plains to find bison herds. The passengers would

Masters of the Hunt   69

disembark, shoot bison, and leave them to rot, except sometimes for the tender meat of the tongue. Although bison populations have rebounded a little from their near extinction, they will never again come close to their former large herd sizes.

Other Animal Population Explosions in North America

There are other examples of animal population explosions that have gone unrecognized (as well as unproven) that may have resulted from the crash of Native populations. I will describe two of them now.

In 1776, botanist William Bartram began a journey through the Southeast of North America, much of it by canoe. When he was on what is today called the Santa Fe River in northern Florida, he was astonished by the large number and size of alligators and fishes. He said the alligators in one spot

> were so abundant, that, if permitted by them, I could walk over any part of the . . . river on their heads, which slowly float and turn about like . . . logs of wood, except when they plunge or shoot forward to beat off their associates, pressing too close to each other, or taking up fish, which continually crowd in upon them from the river and creeks . . . especially the great trout, mudfish, catfish, and the various species of bream. . . . It is astonishing and incredible, perhaps, I may say, to relate [the] unspeakable numbers of fish. . . . Thousands are driven [by the alligators] on shore, where they perish, and rot in banks . . . the stench being intolerable.

Sometimes the alligators attacked him in his canoe. Even on land, when he camped, the alligators approached him. He was frightened by the loud snapping of their jaws and their roars "which made the earth tremble."[12]

This sounds to me like population explosions of fishes and of the alligators that ate them, both of which the Natives hunted and ate. There were still numerous Natives in the vicinity, but when there were more Natives, before smallpox, the hunting and fishing pressure might have been sufficient to prevent these apparent population explosions.

RESURGENCE

Resurgence occurs when a prey population has an explosion following the disappearance of a predator. The resurgent populations are so abundant that they then overrun their resources and suffer a population decline,

70   *Forgotten Landscapes*

even a population crash. The preceding examples included animals for whom Native Americans were an important predator. But resurgence is an ecological process that can occur anytime and anyplace. This is in addition to natural population cycles in which animals become abundant and their populations crash when they overrun their resources or when diseases (which spread more rapidly in crowded populations) kill them, irrespective of predator populations.

Some of the best-known examples of resurgence come from plagues of agricultural pests. Throughout human history, plagues on agricultural crops have threatened millions of people with starvation. One famous example from ancient history is the Ten Plagues of Moses in the Old Testament book of Exodus.[13] Six of the ten plagues sound like pest populations out of control, and the others may have had natural causes. But probably no part of the world and at no time in human history has human agriculture escaped plagues of insect and rodent pests.

I remember old home movies of a locust plague in Oklahoma in the middle of the twentieth century. In ancient times, plagues began and ended without visible explanation. Witnesses often chose religious explanations, not just in ancient times but as recently as the Mormon locust plague in Utah in 1848. And these stories could become the founding myths of a culture. In many cases, these plagues have been caused in part by the elimination of natural predators. Sometimes the return of natural predators (as when nearby seagulls flew in and ate the Mormon locusts) helped to bring the plague to an end.

What can be particularly disturbing is the way the insects coordinate their movements during the plagues. For reasons not well understood, a large number of insects can shift from individual to coordinated movement, in which they all move the same way at the same time and migrate together to new areas. This shift in behavior requires a minimum population density, but nobody knows what this number might be or the actual stimulus of the change.[14]

Starting in the twentieth century, many agricultural plagues have been caused by the overuse of pesticides. Most pesticides kill a wide range of insects, both crop pests and the insect or spider predators that eat them. When the pesticide molecules break down or are washed away, the crop pest populations can rebound much more rapidly than can the slow-growing populations of the predators. This is even more true of predators such as birds, as explained decades ago by Rachel Carson in *Silent Spring*. In this way, the pest populations can explode while the predator populations are still too small to control them. I describe one example below.

The Mouse Plague of Kern County

The southern San Joaquin Valley (pronounced *wa-keen*; chapter 8), which is currently Kern County, seemed like a paradise of riches waiting to be developed by white civilization. The soil was deep and fertile, the remnant of sediments that accumulated at the bottom of Buena Vista Lake, which since the last ice age has largely disappeared. Winters were warm, which allowed the cultivation of high-value tree crops such as oranges, olives, and almonds on the slopes above the former lake bed. A lot of lower-value crops, such as wheat, rye, and maize, were also grown. The summers were long and dry, which made irrigation necessary. All that was needed to unlock the agricultural potential of the Valley was to use the river water that would otherwise have gone into the lake and use it for irrigation.

When there was no longer enough river water for irrigation, farmers pumped groundwater from the shallow aquifers. In addition to the agricultural riches, there were immense oil reserves at the southern end of the Valley. Some of these reserves figured prominently in the oil scandals of the Harding administration. There seemed to be nothing to slow down the phenomenal economic growth of Kern County.

And then, one day in 1926, inhabitants discovered that their fields, houses, and businesses were suddenly overrun with mice. Nobody could keep them out. Mice running over beds kept people awake at night. The mice overran oil fields, camps, and ranches. They ate everything, even, according to one newspaper account, an entire sheep that was confined in its pen. It is estimated that by the time the mice reached Taft, the major oil city of the region, there were a hundred million of them. Desperate oil workers dug trenches for poisoned grain to kill the mice, but the mice bred so rapidly that these efforts at control did not work. The main highway was slushy with dead mice.[15]

Like other animal population explosions, the plague of mice had a complex network of causes. The first cause was that the intensely farmed land had a lot of grains that the mice could eat. The mice had virtually unlimited food at their disposal. The second cause was that the mice were European house mice (*Mus musculus domesticus*), which preferred crops to native plants as a food source, and which had an enormous reproductive potential. A female house mouse can have ten to fourteen litters in a year (which had virtually no winter), each litter with six to eight offspring.

The third cause was that the federal Bureau of Biological Survey, the predecessor of the modern Fish and Wildlife Service, which as we have seen had more interest in than knowledge about wildlife management,

72   *Forgotten Landscapes*

instituted a program in 1924 of paying bounties for people to kill predators such as coyotes and hawks in Kern County. As a result, there were very few predators to eat the mice. The fourth cause was that, despite irrigation, the fields were too dry for the mice to live in them, and they took refuge in the shrinking Buena Vista lake bed. The fifth was that a massive rainfall event in November 1926 started to fill the lake bed. This drove the mice out into the surrounding countryside. The result was what some researchers called "arguably the most dramatic mammalian population eruption reported for North America."[16]

Before the federal pest infestation expert could figure out what to do about the mouse population explosion, something that seemed to the inhabitants like a miracle occurred. Predatory birds such as owls, hawks, and ravens, virtually extirpated from Kern County, flew in from other places and feasted on the mice. By the time another rainstorm event occurred, the following February, many of the mice, in now declining populations, drowned.

Plagues of pests are indicators of a natural world out of balance. In this case, it was massive food availability and the destruction of natural predators. The solution to the problem was the unexpected return of the very predators the people had tried so hard to eradicate.

WHAT WAS NATIVE HUNTING AND FISHING LIKE?

Before European contact, Native hunting and fishing equipment, for an individual hunter, was limited. Nonetheless, the equipment could still be very effective if used skillfully, as was nearly always the case. Bows and arrows, spears, clubs, and even rocks could have deadly accuracy in hunting. Native fishermen used nets and complex fish traps and weirs.

But the real hunting and fishing skills of Native tribes is best seen in groups. It is in this way that Native hunting and fishing skills were at least the equal of, if not superior to, those of Europeans and white Americans. I present here just a few examples of Native group hunting as recorded in the journals of European and white American explorers.

Native Hunters Used Fire as a Tool

In the previous chapter, I gave examples of how Natives used controlled burns to enhance the growth of wild foods that would attract prey animals. But they did more than this. When they located prey animals such as deer, they would light fires around the herd to drive them into a centralized location where it was easier to shoot them. The earliest account of this activity was from an anonymous work about New England between 1607 and 1622.[17] The Native tribes would hunt deer:

> by making of several fires, and setting the Countrey with people, to force them into the Sea . . . and then there are others that attend them in their Boates with Bowes and weapons of severall kindes, wherewith they slay and take at their pleasure.

Father Louis Hennepin, part of René-Robert Cavelier, Sieur de La Salle's expedition in 1683, wrote about Native Americans hunting bison:

> When the Savages discover a great Number of those Beasts together, they likewise assemble their whole Tribe to encompass the Bulls, and then set on fire the dry Herbs about them, except in some places, which they leave free; and therein lay themselves in Ambuscade. The Bulls seeing the flame round about them, run away through those passages where they see no fire; and there fall into the Hands of the Savages, who by these Means will kill sometimes above six-score in a day.

Fire as a tool of hunting was observed as long ago as 1760 by Andrew Burnaby. Anthropologists later recorded and published further accounts about hunting with fire in numerous tribes, from the Algonquins in the Northeast to the Yuchi in the Southeast.

It was not just large game that Natives hunted by means of fire. They also ate a lot of insects, particularly grasshoppers and locusts.[18] Hunting insects individually can mean a lot of work for small bites of food, but Natives would concentrate the insects into a centralized location. Roland B. Dixon, an anthropologist, studied the Maidu tribe of California and described this practice:

> Grasshoppers and locusts were eaten eagerly when they were to be had. The usual method of gathering them was to dig a large, shallow pit in some meadow or flat, and then, by setting fire to the grass on all sides, to drive the insects into the pit. Their wings being burned off by the flames, they were helpless and were thus collected by the bushel. They were then dried as they were. Thus prepared, they were kept for winter food, and were eaten either dry and uncooked or slightly roasted.

Another anthropologist, Catharine Holt, reported that the Shasta tribe of California "ate one kind of grasshopper, a very large form. They

74   *Forgotten Landscapes*

set fire to the grass, thus cooking the grasshoppers which were then dried, pounded, and mixed with grass seed for eating."

Even when the Natives did not use fire, they employed cooperative hunting techniques to drive prey into a centralized location. Deer were trapped in a circle of hunters. Edwin Bryant in 1846 observed this approach used for hunting rabbits. Father Pierre-Jean De Smet observed Rocky Mountain Natives driving a circle of grasshoppers into a prepared hole where they could be easily caught.

Native Hunters Constructed Elaborate Structures to Capture Prey

Samuel de Champlain observed a Native deer hunt in 1615. The hunters took ten days to construct an elaborate corral made of wood stakes pressed together, into which a chute led, its size diminishing as it approached the corral. The deer trapped in the corral were easily killed. Explorers in the West described Native antelope corrals covering up to two acres, with walls twelve feet high constructed from piñon pine branches.[19]

NATIVE AMERICANS INVENTED
WILDLIFE MANAGEMENT

In the previous chapter, I wrote about how fire suppression by white colonists led to unhealthy forests and diminished grasslands. In later chapters, I will write about how Natives practiced sustainable agriculture long before white colonists even thought about it. In both cases, modern American society has begun to incorporate Native practices such as controlled burns and sustainable agriculture. But not often enough.

Another practice that Native Americans invented is wildlife management. Unlike controlled burns and sustainable agriculture, wildlife management has been adopted widely by the public, mainly by hunters, fishermen, and their families. And, in my years of teaching students who grew up in a rural culture saturated with hunting and fishing, I never heard any complaints about limitations on the number, size, and types of fish or game. Unlike every other kind of environmentalism, the rural conservative culture has bought into wildlife management. In some states, Native Americans get exemptions from some of the hunting and fishing rules.

Before European and white American markets turned deer and beavers from essential resources into commodities, Natives hunted only what they needed. Natives were careful even with the most abundant prey, such as passenger pigeons. They did not disturb the adult pigeons, afraid they might disturb the nesting grounds. They ate only the juveniles.[20]

Native wildlife conservation went beyond the merely utilitarian. They saw themselves as part of a spiritual network of all animals. In preparation, sometimes days before a big hunt, they would take morning and evening baths. Ancient Cherokees attributed disease outbreaks to the improperly hunted animal spirits taking revenge.[21]

Native tribes practiced wildlife conservation "from time immemorial," which means that nobody really knows how and when it started. But it went on for centuries, probably for millennia, which means that it vastly predated European and white American wildlife conservation. But wait, you may ask; what about medieval European forest management, restrictions on hunting and logging in such places as Sherwood Forest? In fact, the main purpose of hunting restrictions in Sherwood Forest, and hundreds of other reserves throughout Europe, was not wildlife management, but to keep common people out of the hunting reserves that the nobility wanted only for themselves.[22]

After the final conquest of Native Americans by white military forces, Native wildlife management, including the ethics, was forgotten. Then, the story goes, Aldo Leopold rediscovered it. This is the same man I quoted earlier who shot as many wolves as he could until, one day, he had an almost spiritual revelation. He realized that killing wolves in order to maximize the deer population was not only bad management, but was deeply unethical. His philosophy of nature is captured best in his essay, "Thinking Like a Mountain." He rediscovered not only the practice but the spirituality of Native wildlife management.[23]

Generations of scholars and government researchers have made wildlife management into a science. General ideas do not go very far. Reading Leopold is not a guide to how to respond to a plague of locusts in western North America, or exactly how many deer to allow hunters to kill this year, which will not be the same as last year, or next year. Government conservation officials keep track of the population sizes of game animals, especially deer. They also pay attention to the number of bucks, does, and fawns. From this information, they project the number of deer for which hunting permits can be safely issued. They use the same approach for fishing licenses.

Public lands that are managed for hunters are also the most likely places to have controlled burns. In both of these ways—controlled burns and healthy wildlife populations—public hunting lands may be the best examples of what this continent was like before white conquest.

76   *Forgotten Landscapes*

PLEISTOCENE OVERKILL?

In the parts of the world where humans have coexisted with other large animal species for hundreds of thousands or millions of years, including much of Africa, Europe, and Asia, major animal extinctions did not occur very often. The most likely explanation is that humans and other animals co-evolved in their food gathering, hunting, and reproduction so that humans simply did not overhunt the other animals. This equilibrium continued even as humans gradually developed better and better hunting technologies.

Hunting and Extinction

In the parts of the world where the first contact that large animals had with humans was with technologically sophisticated modern humans, extinctions often followed. The most famous example is the Pleistocene extinction in North America, in which dozens of large animals became extinct at about the same time that the ancestors of Native Americans arrived from Siberia.[24] But it wasn't just here. When modern humans first arrived in Australia, about seventy thousand years ago, there was a major die-off of large Australian animals. When modern humans first arrived in Madagascar, they drove the large moa birds to extinction. In this case, there is archaeological evidence of it. In each of these places, there were no earlier humans whose hunting technology was less sophisticated than arrows and spears.

Extinctions occurred in other places, as well, but at a much lower rate. In Africa, during the same time frame, fewer than 20 percent of the large animal species became extinct, while in North America, it was over 70 percent.[25]

After writing all of these wonderful things about Native conservation ethics, I must now face the possibility that the ancestors of Native Americans swooped in from Siberia and killed so many large animals that extinction became inevitable in their depleted populations. This might not have been as difficult as you might think. The large animals, having never seen humans before, did not run away from the unimpressive-looking hunters, and were easy victims of their arrows and spears.

Everyone admits that hunting was not the only challenge facing large mammal populations. But when Paul Martin first started championing the "Pleistocene Overkill" explanation of North American extinctions, he assigned particular importance to it.[26] This was an extension of the view of early-twentieth-century ecologist Frederic Clements, who believed

Masters of the Hunt 77

humans were primarily disruptors of a natural world that was, otherwise, always moving toward balance.

However, the overkill explanation of the Pleistocene extinctions faces the same problem as attributing the extinction of the passenger pigeon solely to professional white hunters: multiple causation. There can be no doubt that the ancestors of Natives hunted large animals that are now extinct. Projectile points have been found in American mammoth skeletons. And some of the earliest Natives drove bison over a cliff, in one instance killing twenty-three of them.

Aside from this, the evidence for Pleistocene Overkill is slim. First, when the ancestors of Natives first arrived, there were not very many of them. Could a small number of newly arrived Siberians have killed all those animals, even if the animals obediently stood still? Furthermore, even though there are a few mass kill sites, there are not as many as you would expect if the ancestors of Natives went on a killing spree.

Other Possible Explanations

The arrival of the ancestors of Native Americans was certainly not the only thing happening at that time. The climate was also changing, and changing rapidly, as the glaciers melted and retreated. Plants and animals migrated quickly, as I explained briefly in the previous chapter. Rapid climate change is a good time for extinctions to occur, but we must not adopt this as a facile explanation. The climate was changing, but in a way that might have promoted plants and animals rather than depleting them. North America had experienced several previous ice ages followed by periods of rapid warming—about twenty of them—of which four were of major importance. Only after the last one, at the end of the Pleistocene Epoch, did major extinctions occur.

One climatic factor that might have promoted extinctions was fire. As the climate became warmer, and in many places drier, natural fires could spread, destroying enough plant cover that small animals—and some medium-size animals, like deer—could still find plenty to eat, but the large animals could not.

In this case, as in nearly every other situation in nature, we must consider not just multiple causation but also the interaction of more than one factor. In particular, as I explained in the previous chapter, Natives were well known for the landscape-level use of fire. This may not only have temporarily reduced the growth of food plants that large mammals

78   *Forgotten Landscapes*

required, but may also have promoted the spread of fire-dependent habitats that have less food for those mammals. A study using the remains of animals trapped in the famous La Brea Tar Pits in Los Angeles indicated that fires, many or most of them started by Native Americans, transformed the landscape in such a way that it was very productive for humans but unfriendly to large mammals. In particular, the chaparral of Southern California, a shrubby habitat that depends on fire, does not produce enough food for large mammals to live there.[27]

The only thing on which there seems to be widespread agreement is that the Pleistocene extinction was not caused by a meteorite or comet impact that stirred up soot and dust, causing the climate to cool. While there is little doubt that this event occurred and that it caused the climate to become briefly cooler, a lot of people are unconvinced that this cooling thirteen thousand years ago could have influenced extinctions that occurred over the following millennia.[28]

HOW NATIVE AMERICANS LEARNED
TO BE SUSTAINABLE HUNTERS

It may seem incredible that the ancestors of Native Americans, now so famous for wildlife management and careful hunting, could have contributed to the Pleistocene extinctions. But I think it is possible that when they saw the prey they had hunted so freely became extinct, Native Americans invented new social norms and beliefs that transformed their hunting from plunder into management. May all the rest of us learn as well from our mistakes as they did from theirs.

4

FARM OR FOREST?

Agriculture in Prehistoric America

Michael Bloomberg, despite his financial success, was not a popular candidate for president when he ran in 2016. I speculate that one reason for this, unrelated to his progressive politics and mayoral experience, was what many perceived as his arrogance. South Dakota governor Kristi Noem called it "pompous ignorance." The quote to which she and many others referred was something Bloomberg said at Oxford University's business school: "I could teach anybody . . . to be a farmer. . . . You dig a hole, you put a seed in, you put dirt on top, add water, up comes the corn." But to understand information technology, he continued, "You have to have a different skill set, you have to have a lot more gray matter."

I do not need to convince any of my readers that modern agriculture is a lot more complicated than this, particularly since it requires knowledge of the very information technology and marketing that Bloomberg admired. Even centuries ago, in the plow-and-oxen days, agriculture required a lot of intelligence. In some ways, traditional agriculture required more intelligence than modern agriculture. Most agricultural production today is "industrial"; that is, you flatten the landscape, remove all the organisms, and stamp your farm on it. A complex process, but not as complex as trying to farm *with* nature, within its complex limitations.

The way the Natives farmed before first contact with Europeans required a tremendous knowledge of the natural world as well as of the crops. Precisely because Native American agriculture had such limited technology—digging sticks, clamshell hoes, and baskets—it required intelligence to compensate for the limitations. A farm is as much an ecosystem as is a forest or a prairie, and all ecosystems are complex.

80   *Forgotten Landscapes*

THE FIRST AMERICANS

When the ancestors of Native Americans first arrived in North America about fifteen thousand years ago, they moved rapidly throughout both North and South America (chapter 1). The tribes in the advance wave saw empty land and raced to occupy it. At first they were exclusively hunters and gatherers, getting all their sustenance from the wild. The wild landscape was, at that time, filled with food, from large, docile animals to numerous nuts and berries. The small Native American populations felt no pressure to develop any kind of food production system other than to eat what nature directly provided.

But the Natives quickly began transforming their landscapes. As their populations grew, and their technologies developed, they transformed them more and more, first through the controlled use of fire (chapter 2) and then through hunting (chapter 3). But for a long time after their arrival, they continued hunting and gathering, in a fire-modified landscape.

Gathering was itself an advanced technology. To be an efficient gatherer, you cannot just go out in the forest or field and start eating things that look good. Try it sometime. You might recognize the red and black berries on a certain vine (blackberries) as being good to eat, but what about the yellowish berries on that other vine (poison ivy)? A lot of people, even my students who have studied plant identification, are not sure how to recognize poison ivy. Imagine if you were a recent Siberian immigrant in a new continent, where almost every kind of plant that you saw was unfamiliar. You would have to be cautious.

Most berries are edible. That is their function in the food chain. These plant species attract animals to disperse the seeds—for example, when birds eat berries, inside of which are seeds that get transported in their guts to new locations. But many fruits are poisonous, such as the luscious-looking berries of nightshade plants. When the ancestors of Native Americans migrated from Siberia into North and South America, they had to learn all of these new plants. Sometimes they could learn which plants were safe to eat by watching what the animals ate. This was not always a reliable method, however. Deer eat poison ivy leaves, and squirrels eat the fruits. This is because poison ivy is not really poisonous. Its tissues do not have metabolic poisons that damage cells. Instead, the urushiol in the poison ivy tissues provokes a range of allergic responses, severe in some people, mild in others. Humans are almost the only animals that have an allergic reaction to poison ivy. Whenever a tribe arrived at a new place, and decided to stay, they had to learn what to eat and how to eat it.

Wherever they went for the first time, Natives had to use the scientific method to find out what was safe to eat. Eat a little bit, and if you get sick, don't eat any more of it. If you don't get sick, eat a little more. Learning which plants and fungi to eat was a complex process, one that the Europeans, when they arrived after 1492 and rapidly conquered America, were in too much of a hurry to attempt. Poison ivy and its relatives do not grow in Europe. European immigrants had to learn from the Natives how to recognize and avoid poison ivy and nightshade.

Another prominent plant in Eastern North America is pokeweed. It produces large, juicy leaves in the early spring, right when animals are hungry for fresh food. The leaves are toxic because they are filled with oxalate crystals. The only pokeweed leaves that are safe to eat are the very youngest ones, and then only after boiling them in several changes of water. It is very likely that Europeans and white Americans learned how to prepare poke—which does not grow in Europe—because the Natives told them.

LATER NATIVE AMERICANS

As Native populations grew, gathering wild food plants was no longer an easily sustainable way of life. By the time of European contact, Native populations were large enough that the only part of the continent in which nature itself was sufficiently productive to support dense populations of gatherers was the Pacific Northwest, where tribes lived off the bounty of fish (especially salmon) and game, and the abundant nuts and berries of the fire-modified coniferous forest.

In the densely forested eastern part of North America, the Natives gradually invented agriculture as a supplement to gathering. They domesticated some native crop plants, such as sunflowers and sumpweed, both of which had small, edible seeds (figure 4.1).

Meanwhile, the Natives in Mexico developed an agricultural system based on maize.[1] It is difficult to overestimate the importance of maize (which most Americans call corn) to the world food supply. Pre-Columbian Native Americans in Mexico bred this crop, including some of its major variants, such as popcorn, from an unpromising grass called teosinte. Developing maize into a crop on which entire Native societies could be based took a lot of work, intelligence, and time. The Natives in Mexico also developed beans, squash, and tomatoes, some of the major crops they gave to the world.

Figure 4.1. Seeds of sumpweed (*genus Iva*) are edible, but small. North American Natives ate these before the arrival of maize, beans, and squash on trade routes from Mexico. *Author photo.*

When maize, beans, and squash arrived in North America from Mexico along trade routes, long before European contact, many North American Natives quickly recognized their superiority. They gave up sumpweed and adopted these Mexican crops as their new food base. All during the Mississippian era, and long before the time of European contact, Native American societies had so many people that they depended on agriculture, and gathered wild nuts and berries only as a supplement to their agricultural food supply.

Meanwhile, in the dry areas of western North America, the Natives had very little agriculture until some of them developed irrigation (chapter 5). They mostly hunted wild animals which had the ability to eat and digest the tough grasses and desert shrubs. Natives let the wild animals, such as deer and bison, transform the inedible plants into edible meat, which is exactly the same thing that people in the Middle East did by letting sheep and goats graze on inedible plants and eating their meat. The major difference was that the American mammals could not be herded or milked.

Even the Natives who lived in regions with good soil often did not develop extensive agriculture. The tallgrass prairies just west of the Eastern forests had good soil, but it was filled with the sod of grass roots that was nearly impenetrable. Even early white colonizers avoided the rich prairie soil until the invention of the steel plow. Until then, the sod was more suitable for building houses than for raising food.

But many tribes that did not have extensive agriculture grew small gardens. Tribes such as the Caddo, Pawnee, Navajo, Apache, and Wichita grew crops in river basins, where the loose, constantly disturbed soil was not filled with sod and was easier to till. Women usually tilled the gardens, while the men did the hunting. The crops were mostly maize, melons, beans, and squash. Sometimes they would plant the crops, go hunting, then return later to harvest them. Some tribes, such as the Comanche, had no agriculture, but they traded with other tribes for agricultural food items.[2]

POPULATIONS TOO LARGE FOR HUNTING AND GATHERING

The first English to arrive in Massachusetts saw a Native population much too large for hunting and gathering to support. They saw populous villages that planted and harvested maize and other crops unknown in Europe. At first, the outposts started by the Europeans were spectacularly unsuccessful. In some cases, such as Jamestown in 1609, the colonists even resorted to cannibalism—that is, eating their fellow colonists who had just died. They would have starved, but the Natives shared food with them. Many Thanksgiving dinners in America today are organized as charity events for undernourished people. The first Thanksgiving was no exception, but the English colonists were the recipients of the charity.

Besides the tribes that the first English colonists encountered, many other Eastern tribes depended upon agriculture as well. The Cherokee population consisted of about thirty thousand people before European contact, was then reduced by a smallpox plague, then grew back to about what it had been before. The Cherokee tribe had too many people to live without agriculture. Probably every other agricultural tribe had a similar story.

The white conquerors knew this. In 1760, British general Archibald Montgomery attacked the Cherokee town of Keowee in what is now South Carolina as part of a larger war between the Cherokees and the English. Recognizing that the Cherokees depended on their cornfields for survival, Montgomery's forces chased the villagers into the mountains to starve. The troops pulled up the cornstalks one by one.

84   *Forgotten Landscapes*

But this was not the only example. In 1761 Colonel James Grant's British army destroyed seventeen Cherokee towns and 1,400 acres of corn-fields. He drove five thousand Cherokees into the mountains to starve. Some Cherokees survived by eating their horses. Try telling Montgomery or Grant that the Cherokees were savages that lived by gathering nuts in the forest.

By the time of the Revolutionary War, white Americans took up where the British had left off. Colonel Griffith Rutherford destroyed Cherokee crops and villages in 1776. During American colonel Andrew Williamson's campaign that same year, he destroyed Cherokee crops and villages. The Cherokees "fled into the mountains, leaving behind their villages, horses, cattle, dogs, and fowls as well as between forty and fifty thousand bushels of corn and ten to fifteen thousand bushels of potatoes." The Cherokees saw "their towns burned, their corn cut down, and their people driven into the woods to perish." Historian James Mooney, writing in 1900, said:

> More than fifty of their towns had been burned, their orchards cut down, their fields wasted, their cattle and horses killed or driven off, their stores of buckskin and other personal property plundered. Hundreds of their people had been killed or died of starvation or exposure, others were prisoners in the hands of the Americans, and some had been sold into slavery. Those who had escaped were fugitives in the mountains, living upon acorns, chestnuts, and wild game, or were refugees with the British.

And it just kept happening. Says sociologist Russell Thornton, "In 1788, a Cherokee town on the Hiwassee [River] was burned, and many inhabitants were killed in the river while they were trying to escape. . . . In one instance the Americans supposedly beheaded Cherokee women and children who were planting crops."[3]

Clearly, 1788 was long after contact with Europeans, and sometime after the Cherokees had adopted a colonial and American agricultural economy, complete with peach orchards. But the population levels were similar to what they had been before the Cherokees adopted white agriculture.

Even the white Americans who admired the Natives could look right at Native gardens and not recognize them as a food production system. Ralph Waldo Emerson was such an admirer of Natives that he wrote one of his most scathing letters to President Martin Van Buren to protest the Cherokee Trail of Tears. But when he looked at Native agriculture, all he saw was unimproved forest. In his essay "Farming," Emerson wrote that the Native

cannot plough, or fell trees, or drain the rich swamp. He is a poor crea-
ture; he scratches with a sharp stick, lives in a cave or a hutch, has no
road but the trail of the moose or bear; he lives on their flesh when he
can kill one, on roots and fruit when he cannot.

While it is true that even Cherokees could not drain wet ground, they had
certainly moved beyond the hutch-and-sharp-stick stage by the time Emerson wrote his letter to President Van Buren.

The point I am trying to make is this: If anyone today thinks that Native populations were small or ignorant enough to live solely by hunting
and gathering, the colonels and generals who were trying to destroy them
knew that to do so, they needed to destroy Native agriculture.

MONOCULTURES

We modern Americans love our monocultures. A monoculture—*mono* is
Greek for "one"—is an area in which the plants are all totally the same:
just one kind, all the same size, spaced uniformly. While Americans and
Europeans do prefer some diversity in our gardens, modern white Americans expect our farms and lawns to be monocultures, in the following ways.

Uniform. Drive through the cornfields of Iowa or Illinois. Mile after
square mile, you see nothing but corn, all planted as densely as possible to
produce maximum yield, all of it the same age and the same breed, the ears
at all exactly the same height above the ground, all of it ready to be harvested at the same time. In other areas, you see monocultures of soybeans.
There are almost no weeds, and certainly no other crops, within a field.

Toxic. Beautiful they may be, but monocultures are often chemically
toxic. They require a lot of pesticide, herbicide, and chemical fertilizer to
maintain their artificial uniformity, in addition to the fumes from the machinery to establish, maintain, and harvest the crop. And it is not just in
America. The wheat fields of Ukraine, or France, have a similar beautiful
uniformity (figure 4.2). The corn and wheat fields in France do not use
as many chemicals as do those in America, but they still use a lot. In this
way, as in many others, France is ecologically "greener" than America,
but not by much.[4]

In some cases, it is difficult to completely suppress the weeds. In earlier
decades, very few soybean fields in Illinois were without a few proud vel-
vetleaf weeds sprouting far above the bean plants. Today *Abutilon theophrasti*
is a weed, but a century ago it was raised in a monoculture as a fiber crop

Figure 4.2. A cornfield near Strasbourg, Alsace, France looks just like an American cornfield. *Author photo.*

known as Chinese jute. Hundreds of its seeds remain alive but dormant in Midwestern soil. Biotechnology has given modern industrial farmers the last weapon to impose monocultures on farms. Farmers can now pour massive amounts of herbicides on their fields, which may kill the weeds but not the herbicide-resistant crops that are bred to resist them.

Square. We also want our farmlands to be as square as possible. The corn and wheat fields of the American Midwest, and the orange orchards of the San Joaquin Valley (chapter 8), are exactly a mile on each side. The corn and wheat fields of eastern France are similar.

The American practice of forcing the rounded landscape into squares can be traced back at least as far as the Land Ordinance of 1785, in which the Confederated Congress divided much of the land northwest of the newly independent United States into squares. (The Confederated Congress controlled what was to become the United States from 1781 to 1789, when the Constitution was ratified.) The land was west of the Appalachians, north of the Ohio River, and east of the Mississippi River. The "township" squares were seven miles on a side, and were subdivided into one-square-mile plots called "sections." This system of townships and sections is still in use today. The land was assumed to be clear of Na-

Farm or Forest?   87

tive inhabitants, and the American army made sure that it was as clear as possible. The system was the brainchild of Thomas Jefferson, who was otherwise a great admirer of Native cultures. The American army also evicted white "squatters" who had moved into the vacuum of former Native habitation before the survey was completed. Congress sold the sections to individuals or corporations rich enough to afford them. Dividing the land into squares began as a mechanism for sweeping away Native presence from America's new territories.[5]

We also idealize our lawns as monocultures. The quintessential American lawn consists completely of one kind of grass, all of the grasses mowed to exactly the same height. Many rural landowners even in dry regions have huge lawns, which require irrigation and thousands of dollars of mowing. I live on just the amount of money that a single rich landowner spends on his turf. And even small lawns collectively require millions of tons of herbicide to kill the broadleaf weeds. A stock photo image which sometimes appears on billboards may express it best. It shows an area of lawn, perfectly uniform, except for a couple of small weeds. An angry thick-necked white man, who gives the impression of being culturally as well as biologically intolerant of diversity, is yelling and pointing his finger at the weed as if it has insulted him by its presence.

When Old World agriculture began in places such as Mesopotamia, Egypt, and the Indus River Valley, monocultures may have made sense. These people lived primarily in flat, arid lands and their agriculture depended on irrigation. They did not have to cut down forests, but just clear away native dryland vegetation, before planting. The crops would only grow where there was irrigation. Water was supplied to the fields, and drained from them, by a gridwork of little canals. The Native Americans in what is now New Mexico and Arizona also used irrigation (chapter 5), and their fields may have looked like a smaller version of our modern monocultures.

The Middle Eastern pattern of agriculture—square fields of single crops—was not very well suited to the hills of Europe. Nevertheless, this is what the Middle Eastern people planted when they arrived in Europe. Genetic studies have shown that farmers, not just farming, migrated into Europe thousands of years ago.[6] The first European farmers cut down forests and imposed square fields on a rounded landscape. The square field mind-set—and the crop species themselves (such as wheat)—came from the Middle East. Later, when many Europeans migrated to eastern North America, they imposed the European monoculture mind-set that was even less appropriate in America than it had been in Europe.

88   *Forgotten Landscapes*

POLYCULTURES AND AGROFORESTRY

We all think we can tell a farm apart from a forest at a single glance. But with the kind of agriculture that the Natives of eastern North America had, the distinction would not have been so clear. A Native American farm, except for some of them near large cities such as Cahokia, would seem to us like a forest, until you looked at it closely and started asking questions about what you saw. Most Native American farms in the East, during the time of the Mississippian culture and continuing up through European contact, were not so much agriculture as *agroforestry*.

The fields in which Natives raised crops looked like gardens, in two important ways.[7] First, unlike modern monocultures in which a single kind of crop covers a large area, Native agriculture was built on *polycultures* in which several crop species lived together. Second, rather than cutting down a whole forest to plant agricultural fields, the Natives planted their crops in spaces between trees, a practice today called agroforestry. Polyculture and agroforestry—this was the key to the success of Native agriculture that allowed it to support large populations.

Modern agricultural research has rediscovered the advantages of polycultures and agroforestry.[8] A lot of agricultural research, such as that conducted at The Land Institute in Kansas, has abundantly demonstrated that polyculture has numerous advantages over monoculture. Polyculture and agroforestry have not just ecological advantages but economic ones as well, as described here.

In a polyculture, crops use different resources. In a garden–type polyculture, the different kinds of crops compete less directly than in a monoculture. The roots of different crops may grow at different depths, and the stems and leaves at different heights above the soil. One kind of crop uses resources differently from the other kinds. In fact, leguminous crops such as beans have little chambers inside their roots in which bacteria transform gaseous nitrogen from the soil air spaces into nitrogen fertilizers such as ammonia, nitrites, and nitrates. If the farmer grows corn, beans, and squash together, the beans do not compete with the others for nitrogen.

In a polyculture, diseases and pests spread more slowly. Pests frequently specialize on just one kind of crop plant. They often spread rapidly through monocultures, since for any given kind of pest, a monoculture represents a limitless feast, with no inedible plants blocking its spread. Crop diseases spread rapidly through monocultures for the same reason.

A polyculture of crops may cause less soil erosion. A polyculture that contains perennial crops—crops that are not plowed up and replanted every year—also has less soil erosion than a monoculture of crops that live for only one year. The farmer does not have to spray as much pesticide or use as much fuel and equipment for plowing. As a matter of fact, the soil often builds up because leaf residue accumulates.

Before European contact, Native agriculture was based on digging sticks and on seed broadcast, not on plows. But since the farmers used digging sticks to plant corn, and they broadcast the bean and squash seeds, the work was not any more onerous than a plow would have been. The digging stick would allow corn to be planted without disturbing the upper soil levels, and thus resulted in far less soil erosion.

Polyculture is therefore the scientifically and ecologically most advanced kind of agriculture, more so than the monocultures that we call modern. Our modern economic system pays for massive inputs of chemical fertilizers, pesticides, and machinery, which makes profits for agribusiness corporations while it keeps the farmers in debt. Much of the money comes from government inputs of tax funds.

Like many Indigenous peoples around the world, Native Americans invented polyculture in America before modern agricultural researchers did. But they didn't realize they were doing it. They weren't thinking, *Our mixture of crops uses resources more efficiently, slows the spread of diseases, and saves the soil.* They were stewards of the land, but they probably didn't think about it. They just grew their crops this way because this is the way they saw nature producing food. For people with digging sticks and clamshell hoes, it was a lot less work. When Natives saw Europeans working hard with plows and oxen to produce monoculture fields, they were probably puzzled about why white farmers wanted to do so much work out in the hot sun after having cut down the trees.

In some cases, Native agriculture would build up the soil even more. Elementary school students learn the story of how Squanto, almost the lone survivor from a smallpox plague that killed most of his tribe in the Powhatan Confederacy, taught the English Pilgrims in New England how to drop food scraps into each hole where corn was planted. This practice would leave agricultural soil more fertile than it started out.

Perhaps the most famous Native agricultural story of all is the legend of the Three Sisters.[9] It is widespread among Native tribes. It probably originated from the milpa system of polyculture practiced for centuries in Mexico and Central America. It may have diffused northward from Mexico along with the original crops themselves, spreading throughout eastern North America by about 1000 CE.

90   *Forgotten Landscapes*

The Three Sisters are maize, beans, and squash. The farmer plants the corn and broadcasts bean and squash seed. The bean and squash leaves grow all over the ground, shading out the weeds that would otherwise grow, while producing an edible yield. The Natives could have easily seen the bean and squash leaves shading out the weeds. But they could not have known that the beans had nitrogen-fixing bacteria in their roots that produced fertilizers that would benefit all three species. That is, the Three Sisters agricultural system was even better than the Natives thought it was. And it worked in most places.

The Three Sisters was not merely a food production system. It was also foundational to the cultures and religions of many of the agricultural Native tribes. It was sacred knowledge given to them by the same Spirit that gave them the crops themselves.

One of the advantages of the Three Sisters system is that the corn grows up in the air where the squash and beans cannot. The corn acts as a trellis for the beans. This part of the process does not always work, since in some places the corn plants may not be strong enough to hold up the other plants, which start growing earlier than the corn. Nevertheless, in many cases, the tribes made this system work beautifully, and profitably (figure 4.3).

Another essential component of polyculture is diversity *within* each type of crop. Natives did not just raise maize; they raised lots of different kinds of maize. They had to think ahead, in their planting, to prevent the different breeds from blending together. In their Appalachian homeland, the Cherokees planted different breeds of maize in different mountain valleys. They may have had a vague idea that corn pollen, which blows in the wind, could cause the different varieties to interbreed. They also raised lots of different varieties of beans and squash.

Crop varieties (which today we often call "heirloom varieties") are an important part of a tribe's heritage. In moving to a new location, a tribe would take seeds of their favored crop varieties with them. This did not happen with the Cherokees, however. The American army came and uprooted the people from their homes and fields, giving them hardly enough time to pack their belongings, much less their seeds. The army assured them they would be given food during the trip and when they arrived in the West. If Elizabeth (see the introduction) asked herself what she would plant when she got a farm in the West, she probably would have thought the army would provide seeds. And they did. But they were not the traditional Cherokee seeds. Any traditional maize seeds that the Cherokees might have brought with them probably interbred with the corn varieties given to them by the American suppliers.

Farm or Forest? 91

Figure 4.3. Reconstructed Cherokee polyculture at New Echota, Georgia. The Three Sisters are maize, beans, and melons. The gourds are houses for the birds that eat insect pests. *Author photo.*

Thus, very few traditional Cherokee crop varieties survived. Cherokee ethnobiologist Pat Gwin told me of the immense amount of work he and his associates have done to even partially reconstruct the Cherokee traditional varieties of crop plants. Here follows the story of how he almost didn't save the traditional Cherokee crop plants.[10]

In 2005, Pat Gwin started from scratch. He was in charge of Cherokee ethnobotanical resources at the tribal headquarters in Tahlequah, in northeastern Oklahoma. He heard about a man named Carl Barnes who lived out in far western Oklahoma. Carl had a huge collection of traditional Native seeds. His collection was disorganized, consisting of shelves and shelves of glass jars in his garage filled with seeds collected from all over North and South America. Carl was losing his memory and had not kept good notes,

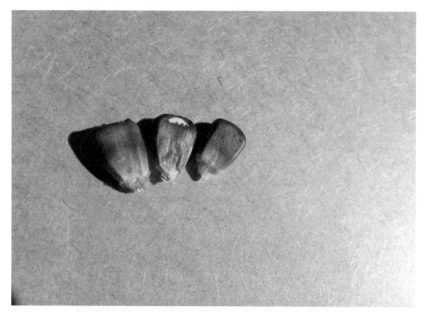

Figure 4.4. The Cherokee Nation recovered its traditional white eagle by breeding from the few remaining seed sources. The white portion resembles the profile of a flying eagle. *Author photo.*

so Pat could not always link samples to any particular place of origin. Some of the specimens were over fifty years old and had very poor germination. Carl gave Pat four Cherokee corn breeds: white eagle, and three flour corns. The white eagle corn, named for the color pattern on the seeds, is the emblematic Cherokee heirloom crop (figure 4.4). Carl's collection has found a new home with the Onondaga Nation in New York, where the tribal members are growing many of the strains of these crops.[11]

Pat also contacted the Eastern Band of the Cherokees in North Carolina, descendants of a few Cherokees who had evaded the Trail of Tears. The USDA contractors who worked with the Eastern Band had lots of traditional seeds, but as was the case with Carl Barnes, their collection was not organized. They brought the seeds to Pat. These included, in addition to some of the same four corn varieties he had gotten from Barnes, a mixture of many unidentified varieties of beans and squashes.

Pat had to try to reverse the effects of cross-breeding among the Cherokee heirloom plants he got from Carl's collection and from the Eastern Band. He bred out the traits from each corn breed that belonged to a different breed. Most of the white-eagle corn that he started with didn't even have the white eagle on the seeds. He had to breed it back into existence.

Pat continued to get seeds from unexpected sources. He did not go out and gather seeds from Oklahoma Cherokees, since most or all traditional varieties of seeds had been lost on the Trail. But there was one old woman in 2021 who had what she had been told was a traditional Cherokee pumpkin variety, and it turns out she was probably right. There was also a museum in Minneapolis that had a Cherokee tobacco pouch, which had some seeds down in the lining. Pat got nine of the seeds to germinate. This was the source of all the Cherokee heirloom tobacco.

In 2020, the Cherokee Nation of Oklahoma sent the reconstructed traditional varieties of seeds to the arctic seed vault in Svalbard, north of Norway, where the permanently frozen ground preserves traditional crop seed varieties from all over the world. The Cherokee seeds are not available for commercial breeding without tribal permission.

Genetic variety within crop species is an essential component of polyculture and, as Pat Gwin discovered, it can be easily lost. All over the world, traditional farmers are giving up their old heirloom varieties of crop and garden plants so that they can plant high-yield modern varieties. One can hardly blame them if they were faced with hunger. But, as Kent Whealy (who founded Seed Savers Exchange, a repository of heirloom varieties of minor garden crops) said, they are literally eating their future. It is from these varieties that modern breeders can obtain the genes for resistance to new diseases or adaptation to poor growing conditions, such as what one may find in marginal agricultural areas or in a globally warmed world.

Another characteristic of Native American agriculture that Europeans did not recognize was the practice of allowing the crops to reseed themselves. Native tribes of the Great Lakes region harvested wild rice (*Zizania aquatica*), which is not the same as Asian rice. It grows naturally in almost-pure stands in shallow water. Natives of several tribes harvest the grains from boats. The plants have to grow from seeds every year. But the Natives also harvest the grains in a manner that modern farmers would consider inefficient—that is, they allow some of the seeds to fall into the water and germinate the next year. A well-meaning scientific advisor devised a system by which modern Natives could gather more of the wild rice seeds with the same amount of work. The Natives were not very interested in the system, because the seed loss was a deliberate part of their encouragement of crop regeneration. A stand of wild rice may not seem like a farm, but in this sense, it is.[12]

California natives would beat seeds of wild herbaceous food plants into baskets, letting many of them drop on the ground, replenishing the food plant population in the same way the Great Lakes Natives did, and

94   *Forgotten Landscapes*

do, with wild rice. Both the wild rice and the California wild food plant examples are a kind of agriculture that, at first glance, seems like nothing more than gathering.

As I explained in chapter 2, California Natives gathered the underground corms of brodiaea, a relative of the onion, from the grasslands to eat, but also to replant in new locations where they would spread. They also gathered their preferred seeds in baskets, and then burned the grasslands, after which they would scatter some of the seeds. Depending on your definition of agriculture, this might qualify.

Agroforestry is like the kinds of polyculture described above, except that one or more of the crops are trees that grow around the fields. Agroforestry, like polyculture, is a kind of agriculture that doesn't look like it. Native American agroforestry in the eastern part of the continent looked like the forest in which it was planted. The Natives were comfortable with plants growing, and producing edible products, surrounded by trees which also produced edible or otherwise useful products (chapter 6).

Agroforestry is even better than polyculture. When a crop field is surrounded by trees, birds can hide in the branches and swoop out to eat insect pests on the crops. Little birds will be less likely to do this if they have to fly a long way from distant forest trees, exposing themselves to their own predators, such as hawks. In addition, the trees protect the crops from excess wind and storm damage. Nobel Peace Prize winner Wangari Maathai championed agroforestry in Africa.[13]

One of the biggest advantages of agroforestry is the benefits conferred by the trees themselves, such as cool shade.

Cool Shade

Another problem confronting all people (at least before the era of central HVAC) is how to stay cool during the summer and warm in the winter. Trees help us to stay cool in the summer, not just because of their shade, but because of their *cool* shade. The evaporation of water from the leaves, a process called transpiration, cools the leaves and therefore also the shade underneath them. The heat from the sunlight is not destroyed—energy is never destroyed—but it rises with the water vapor into the upper air, away from where we are trying to stay cool. In this way, trees can function as air conditioners. During the summer, trees cannot fully take the place of air conditioners, but they can greatly reduce the heat load for which our air conditioners, buzzing like angry wasps, are called on to address.

In my book *Green Planet*, I described white soldiers at a fort around 1800.[14] "Sweating in the hot Georgia sun, the soldiers seek shade under canvas and log. They sit and curse the heat, hanging out their tongues in anticipation of something to drink." Meanwhile, in a nearby Cherokee village, the inhabitants sat in the cool shade of trees inside of their stockade. The point I was trying to make here is that simply by not cutting the trees down, the Cherokees were able to enjoy the benefits of cool shade inside their villages.

Building engineers know that trees provide free air-conditioning—that is, if you have the right kinds of trees at the right places, along with plenty of water. The Cherokees and other Natives already knew this. This idea did not have to be rediscovered. The benefits of fire management, game management, polyculture, and agroforestry also did not need to be rediscovered. I wonder how many thousands or millions of dollars are still being spent today on environmental research to just rediscover what Native Americans already knew.

In the winter, deciduous trees drop their leaves and allow sunlight to warm us up. The bare branches also slow down the winter wind that might otherwise blow away the warmth. As with air-conditioning, trees cannot take the place of heaters, but they can help to reduce the energy load the heaters would otherwise use.

Unfortunately, our modern American response to a heat wave is to crank up the air-conditioning, which means power plants need to discharge even more carbon dioxide, which causes more global warming. It is a buildup of cause and effect that we cannot win. A newscaster noted that it is inhumane to expect people to live without air-conditioning. This assertion is questionable. Very few people in France, where I now live, have air-conditioning. Local American utilities and governments have in very few cases taken the initiative to encourage people to be moderate in controlling indoor temperature. Democratic president Jimmy Carter in 1979 tried to encourage people to turn their thermostats down to 68 degrees in the winter, and many people laughed at him for it.

Costs and Benefits of Agroforestry

There are some costs associated with agroforestry, of course, as with everything. The trees partially shade the crops, potentially reducing their photosynthesis and growth. On hot, dry days, however, the partial shade can be welcome, and the crops may grow better than they would in full sunlight.

96   *Forgotten Landscapes*

Although polyculture and agroforestry presented fewer pest problems than monoculture, pests could still overwhelm Native farmers. Insects still managed to find the crops, especially insects like grasshoppers that could eat many kinds of crops; to them, a polyculture was as much of a limitless banquet as a monoculture would have been. And the birds in the trees could not eat all of the pests. Often the farmers had no alternative but to manually kill the insect pests. Early-twentieth-century Native farmers, like my Cherokee grandfather Edd, had no access to pesticides. But he got his kids, including my mother, to pick the insects off the plants and throw them into a big pot of hot water to die.

And then there were the crows. They would eat newly sprouted corn, and then, after the corn plant got too large to pluck, they would wait until the corn was ripe before eating it. Crows, among the smartest of birds, can find corn even when it is mixed up with other crops. Eyewitness accounts of eighteenth-century Cherokee and other Native cornfields describe how women and children would sit on wooden platforms and watch for the crows. When they saw them, they would raise a racket to scare them away, sort of like living scarecrows.

CROP ROTATION AND FALLOW

Another form of polyculture is what we now call *crop rotation*. It is crop diversity in time rather than in space. In one year, one crop may dominate a field, and begin depleting the soil in a certain way. In a subsequent year, a different crop may dominate the field, and begin depleting the soil in a different way. During the second year, the field may begin to recover from the damage caused by the first crop. This is especially true if one of the crops is a legume, which enriches the soil with nitrogen fertilizer. Even without knowing the scientific basis, many cultures independently discovered the benefits of crop rotation.

A related concept is called *shifting agriculture*. After growing crops in a field for a few years, and witnessing a decline in fertility, the inhabitants may simply abandon it. This is similar to crop rotation, but the rotation is crops first and then weeds, which may regenerate the soil in a way similar to a complementary crop. The soil regeneration is even better when the farmers burn the weeds, as the Native farmers did. After a few years, the abandoned field can be cultivated again. As with crop rotation, many cultures have independently discovered the benefits of shifting agriculture.[15]

As we have seen, polyculture was elevated to the level of religious practice in Native cultures. But it wasn't just in America. Shifting agriculture is a commandment in chapter 26 of Leviticus in the Old Testament of the Bible. This chapter records instructions (supposedly from God Himself) to cultivate the fields and eat their produce for six years, but the seventh year is the "Sabbath of the fields," during which time the field is supposed to "rest."

There is no evidence that the Israelites ever practiced the Sabbath of the fields. According to the biblical timeline, 490 years of Israelite occupation of Palestine accrued a debt to the land of seventy years. According to the second book of Chronicles, chapter 36, not only was seventy years the exact amount of time the Israelites were to remain conquered by the Babylonians, but this was the reason. "And so the land enjoyed its rest," wrote the chronicler: seventy years all at once, after the country's collapse, rather than seventy years spaced out as part of a sustainable economy.

Crop rotation and shifting agriculture were extremely important in maintaining soil fertility. It has been incompletely and inadequately replaced in modern agriculture by the use of chemical fertilizers. Crop rotation and shifting agriculture also help to control pest populations that can build up year after year, but which may crash when the crop upon which the pest depends is not planted in that location.

THE END

As the above examples indicate, there is no food production problem to which Natives did not have a solution, at least on a small scale.

An unaltered forest, however, would not be a good place to practice polyculture agroforestry. It would be choked with shrubs, vines, and small trees. Therefore, near the villages, the Native farmers cleared away the smaller plants, often using fire (chapter 2). They did not clear away the large trees, mostly because it would have been just too much work. They cut down large trees only when they needed to make canoes or longhouses out of the trunks.

By removing the undergrowth, Native farmers produced parklike woodlands through which travel was almost unobstructed. The Europeans were amazed at these "wild" American forests through which they could ride their horses easily. They considered these forests to be a creation of God, available for them to conquer. But these parklike forests were created by Native Americans.

By the time Europeans had penetrated further into the continent, epidemics had spread, killing many Natives, and many of the villages and their farms had vanished into the undergrowth.

5

THE BLESSINGS OF WATER

Irrigation in Prehistoric America

We grow up learning that water is invisible and has no taste. But this is not quite true. When water bends beams of light, it appears blue. And while you do not have taste buds or scent receptors that respond directly to water, you can subconsciously react to the presence of water. Mice can use their acid taste buds to detect the presence of water. And the response is a pleasant one. Researchers refer to the "hedonic value of water-evoked responses," which means the mice not only detected but enjoyed the sensation of water.[1]

I have had this experience and so, possibly, have you. While participating in high school cross-country, I once ran past someone's sprinklers in an arid area of California and imagined I could taste the water—a lovely feeling. Water is not just something we need, but something we deeply and subconsciously enjoy.

Humans can eat lots of different kinds of food. And as for energy, there are lots of different sources, from biomass to solar to fossil fuels. But there is no substitute for water. We have to have it, individually and collectively, not just for our bodies but for our entire economy. This is the reason that there is a tight correlation between human population density (high in the east, low in the west, of North America) and water availability, today just as it was thousands of years ago. The western part of the continent is pretty dry, sometimes severely so, and anyone who intends to live there has to find sources of water. Cherokees in the East could trust the rain to water their crops, but the desert tribes needed irrigation and other forms of water management in order to have a civilization. Water is a concentrated and irreplaceable blessing.

Native Americans and related peoples were famous for being able to make a living in the most challenging habitats, from the Inuits on the frozen tundra to the Fuegians at the icy tip of South America. One of the most challenging habitats is the southwest American desert. Water, and not much of it, is briefly available as light rains in the winter and spectacular floods in late summer. Humans have long had many creative and inventive ways to find food and water. And all over the world they found ways to get the water when and where they wanted it.

One of these ways was to encourage rapid and brief flows of water to percolate down into the soil where plant roots could get it during prolonged periods of arid conditions. Another way was to build irrigation canals to transport water from where it is usually available to places where it is not. There is no clear dividing line between these two ways of managing water. In the southwestern deserts, Native American tribes did both.

PAIUTE CANALS

The Paiute tribe of Native Americans lived in some of the hottest and driest regions of the Southwest—places so hot and dry that it is surprising a jackrabbit could survive there (figure 5.1). But they managed to do it.

Figure 5.1. Owens Valley lies in the rain shadow of the Sierra Nevada. The rain and snow stay largely on the western, Pacific slope, on the other side of the mountains as seen from this viewpoint. On the east side, there are very dry deserts, including not only Owens Valley but also Death Valley. Mt. Whitney is one of the high peaks in the center. *Author photo.*

They knew where to find plants that were nutritious, such as the seeds of many species of wild grasses and pigweeds, and tubers such as nutsedge and wild hyacinth.[2] These plants grew in the tiny spaces where there was adequate water, usually in arroyos and along river margins such as the Colorado and Owens Rivers. The arroyos filled with floodwaters after heavy summer rains. The rivers ran all year because of snowmelt from the Rocky Mountains (into the Colorado River) and the Sierra Nevada (into the Owens River).

But the Paiute did not merely gather wild plants. Between approximately 600 and 1000 CE, the Paiutes of Owens Valley, on the dry side of some of the highest mountains in North America, began to build an ever more extensive network of earthen canals that extended several miles from a dam on Bishop Creek to two immense fields, each about two square miles in extent.[3] Here they transplanted and raised the same plants they had previously gathered from the wild.

It is difficult to see any remnants of the canals today unless you know what to look for. On the several occasions I visited Owens Valley, I did not suspect their existence and did not see them. The person who first revealed the existence of these irrigation canals was not someone you might expect. You may remember Julian Steward from chapter 2 as the archrival of Omer Stewart. With a white bias typical of scholars of the early twentieth century, Steward considered Native Americans to be savages who could not possibly have used fire as a landscape management tool. Nor could they have had agriculture. But the evidence for pre-contact Paiute irrigation canals was strong enough that he admitted it in 1930. However, he said, it couldn't really be considered agriculture, since the Paiutes were raising wild plants, not crops. By a narrow definition of agriculture, based on plants that have been bred to grow on farms, he was right. So he called the Paiute phenomenon "irrigation without agriculture."[4]

WELCOME TO CHACO CANYON

When we think of Native tribes in the American Southwest, the Navajo tribe comes quickly to mind, even though they were not there until about 1600. When they and the Apache arrived from the North, they found abandoned cities and even skeletons of not just previous inhabitants, but an earlier civilization. *Anasazi* is a Navajo word for this earlier civilization. We do not know what the Anasazi called themselves. One of their urban complexes is referred to today as Chaco Canyon in northwestern New

Figure 5.2. Some of the reconstructed walls in Chaco Canyon, New Mexico, where there was once a thriving Native civilization in the desert. *Author photo.*

Mexico. One of their last outposts was the cliff dwellings of Mesa Verde in southwestern Colorado.

It is easy for even the most observant traveler to be totally unaware of Chaco Canyon. When I went there in 2006, I had to drive for miles on a washboard-rutted dirt road. What I then saw was the sandstone walls of buildings, with tree-trunk support beams but without roofs (figure 5.2). When the ruins of Chaco Canyon were discovered by modern scholars, many of the stones were in disarray. Archaeologists worked very hard to put the walls and the wooden support beams back together.

A less promising location for a civilization is hard to imagine. The American archaeologists who studied it said it was the worst possible location for a desert resort, and one of the main reasons was the virtual absence of water sources. Great civilizations have grown in arid lands—Mesopotamia and Egypt come immediately to mind—but only where there was a large and reliable source of water, such as the Tigris, Euphrates, and Nile Rivers. But the only watercourse in Chaco Canyon is a small wash that is dry except for periods of rain, and then only in good years. The people who built these cities had to make a deliberate decision to locate them here.

The peak period of occupation of Chaco Canyon was 900 to 1150 CE. Studies of tree rings show that most of the timbers were cut between 1030 and 1120 CE. How do archaeologists determine these dates from tree rings?

Tree Ring Chronology

As I briefly explained in chapter 2, trees produce a new ring of wood each year. They do so because the earlier layers have become damaged and clogged, reducing their ability to transport water from the roots up to the leaves. The older rings retain a little of their function, but a tree relies mostly on its newest layer of wood. You can usually tell how old a tree is by counting inward through the wood layers, usually from a cylindrical core. On a dead tree trunk, you can tell how long the tree lived in the same manner.

Some tree species have thicker wood layers than others. The tree produces wood primarily in the spring, and hardly at all during the winter. Each ring represents one year of growth. In trees that live in harsh conditions—cold, or dry—the layers are very thin. Trees that live in warm, humid environments have thicker layers. In addition, trees that grow rapidly have thick layers whose large vessels provide a lot of water for rapid growth, while trees that grow slowly have thin layers of wood. In tropical rainforests, which have neither winters nor dry seasons, the trees have no rings at all. Among the thinnest wood layers in the world are those of the extremely slow-growing and extremely old bristlecone pines of the mountains of eastern California (not far, in fact, from where the Paiutes live) (figure 5.3).

Figure 5.3. Some of the thinnest tree rings in the world are found in bristlecone pines (*Pinus longeava*) of the dry mountains of eastern California. The extremely thin layers are best visible right underneath the dime. The large cracks came from wood deformation between some of the rings, or along the xylem rays. A trunk two meters thick can have thousands of wood rings. *Author photo.*

104   *Forgotten Landscapes*

Each year's ring of wood is usually thicker or thinner than any of the other rings in the same tree trunk, for the same reasons. A dry year or a cold year can result in a thin layer of wood, and a wet or warm year can result in a thick layer. The tree ring layers in a trunk found at an archaeological site are, therefore, a record of climate conditions extending from the year the tree was big enough to produce wood layers until the year it died. The climate conditions would be the same throughout a generalized area, such as northeastern New Mexico, and in all large tree species. By comparing the thick and thin layers of wood in an archaeological specimen with those in a living tree, you can line up the pattern (which looks like a bar code) of the older layers of living trees with the younger layers of the specimen. This is how archaeologists can determine the year in which a tree was cut down, or, in the case of Chaco Canyon, pulled down (since the Anasazi Natives had no tools capable of chopping down trees).

An Urban Center in the Desert

During this brief century of construction, and during the approximately three centuries of occupation, Chaco Canyon was an amazing place. The most obvious evidence is the buildings themselves.

Chaco Canyon had at least nine multistory buildings, some originally with hundreds of rooms. The walls were constructed from blocks of sandstone tightly fitted together (figure 5.4).[5] The walls could bear the weight of upper stories because each one had two sandstone faces with rubble between them, making them very thick. Wooden beams supported the roofs and the floors, which were also made from wood.[6] The buildings were like apartments, with shared walls.

Perhaps the most amazing aspect of the buildings in Chaco Canyon is that they collectively required the wood of 240,000 trees, most of which were ponderosa pines and Douglas firs.[7] The local trees, such as junipers and piñon pines, are too small to be used for roofs and floors of such large buildings. The early, smaller buildings at Chaco Canyon used timbers of piñon pine, which were available locally at the time the first buildings were constructed.[8]

Then as now, the closest sources for the ponderosas and Douglas firs used in the larger buildings are mountains that are at least fifty miles away.[9] The Native builders had no metal tools or draft animals. They had to light fires at the tree bases and pull the trees over, then carry them by hand to Chaco Canyon. This was a massive and deliberate effort.

Figure 5.4. Detail of one of the walls of a Chaco Canyon building. The wall faces were made of large and small pieces of sandstone. Tree trunks held floors and roofs in place. *Author photo.*

There is even evidence that Chaco Canyon, like other Native and Old World civilizations, made use of advanced astronomy. The Anasazi altered certain rock formations so that sunlight would strike certain locations on the rock at the solstices, and the buildings as a whole were aligned east

106　*Forgotten Landscapes*

and west with a precision that seems beyond chance. Whether the Anasazi invented their own astronomy, or they got it from Mexico, which already had it at the time, cannot be determined.

But Mexico is where the Chacoans got their agriculture. Like most North American tribes (except the Paiute), the Anasazi grew maize, beans, and squash before European contact. There is simply not enough water, nor was there apparently during the phase of peak occupation, to raise substantial amounts of these crops. Some kind of water management was necessary. The Anasazi grew crops on mesa tops surrounded by rock walls that slowed down the rain enough to allow water to infiltrate the soil. They also built dams in arroyos, allowing water to be diverted to fields when it was briefly available. The inhabitants supplemented their agricultural crops by gathering wild cactus fruits, mesquite pods, and agave.

The people also needed protein. Early in the Chacoan civilization, the nutritious seeds of piñon pines might have been readily available, as venison would have been. As piñon nuts and deer became scarce, the Anasazi, like most other Natives, also raised domesticated turkeys.

But how much crop production was necessary? Some estimates of the Chaco Canyon peak population are close to 2,300 people. There was certainly enough room for them in the buildings. Some scholars have estimated that there was no way the local resources, even at the time, could have fed 2,300 people with agricultural produce and, certainly, with protein from hunting.[10] This does not mean, however, that there was a year-round population of 2,300 people. Far fewer people might have lived there permanently, with large numbers of religious pilgrims converging during brief periods of religious celebration. This was, after all, what seemed to have happened at Spiro, Oklahoma, as I mentioned in chapter 1. The pilgrims might have even been required to bring their own food.

One piece of evidence that Chaco Canyon was a focus of religious activity is the overwhelmingly large number of kivas. Modern Pueblo tribes, and presumably their Anasazi forebears also, had religious ceremonies in underground stone-lined rooms. There was about one kiva for every thirty rooms (figure 5.5).

As in Cahokia and Spiro (chapter 1), and indeed, in Mexico, the Anasazi religious leaders were very powerful and wealthy. Their skeletons showed greater height and health than the average inhabitant. Among their wealthy grave items was a necklace containing two thousand turquoise beads. Even if the ordinary people had barely enough to eat, the rich imported luxury goods from Mexico. They reinforced their image of power

Figure 5.5. A reconstructed kiva at Chaco Canyon. *Author photo.*

not only by having workers construct many large buildings, but also the long and broad boulevards connecting them.

The Last Days of the Anasazi

Nobody is entirely sure why the Chaco Canyon civilization collapsed, except that it had nothing directly or indirectly to do with the arrival of Europeans. By the time the first Spaniards arrived in what they called New

108   *Forgotten Landscapes*

Mexico, bringing Old World diseases, the Anasazi civilization was already gone. The collapse was almost certainly connected with a prolonged period of drought, even by arid land standards. The drought of 1130–1190 CE was followed by climatic uncertainty, in which some years had enough water for crops, and some did not. During these years, almost no amount of water management, using the technology of the time, would have been enough to feed the people of the entire region, whether they were residents of Chaco Canyon or not.[11]

In addition to the drought, there was also environmental degradation. The use of piñon trunks for the construction of early buildings was probably not enough to make the piñon-juniper woodlands disappear, but these piñons and junipers were apparently the major source of firewood, to keep the inhabitants warm in the sometimes brutal winters and to allow them to cook their food.[12] There are virtually no piñons or junipers in Chaco Canyon today. But plant material preserved in packrat burrows shows that piñons and junipers were common during early Chacoan times.[13]

Most of the thirteenth century was a time of continual warfare. While there is no evidence of big battles, inhabitants abandoned some of the settlements quickly, leaving behind utensils and clothing. Some archaeologists have suggested that cannibalism occurred. Evidence of cannibalism includes long bones (such as leg bones) broken to extract the marrow; burned bones; cut marks on bones produced by stone knives; abrasions produced by striking bones with rocks; and even a sheen left on bones when they are boiled for a long time in a pot. In some cases, the sheen contained a human version of myoglobin, a muscle protein.[14]

The major Anasazi response to deforestation, drought, and warfare was to migrate to new locations. One of these new locations was Mesa Verde, in southwestern Colorado, preserved today in a national park. Here we find large buildings constructed of stone and wooden beams obtained locally. The last of these buildings was constructed around 1286 CE, a century after the start of Chaco Canyon depopulation. The drought was not as bad here as it was in Chaco Canyon. For a while, at least, there was enough water to permit the people to raise maize, the intact seven-hundred-year-old cobs of which can still be found in one of the storerooms.

The major and obvious difference between Chaco Canyon and Mesa Verde is that the latter buildings are called "cliff dwellings" because they are constructed within caves on the sides of mountains (figure 5.6). Even today they are nearly inaccessible except where tourist pathways have been constructed. The inhabitants (and warriors) could get in and out only by ladders constructed of wood and rope. The buildings included numerous

Figure 5.6. One of the Anasazi cliff dwellings at Mesa Verde National Park in southwestern Colorado. You can still see the smudges from extensive fires on the roof of the cliff. *Author photo.*

110   *Forgotten Landscapes*

defensive walls. If a cliff city was under attack, the inhabitants could remove the ladders. It was an immense effort, devoted entirely to defense. Although there were kivas, they were not centers of religious pilgrimage.

Even the best defenses could not protect the last Anasazi outposts from collapse. This time, unlike Chaco Canyon, deforestation probably did play a major role in the collapse itself. The winter temperatures could be very cold, and sunlight was just not enough to heat the dwellings. The inhabitants might have stripped the piñon–juniper forests for firewood. A park service employee told us that, in the last days of Mesa Verde, "there probably wasn't a stick of wood on these mountains."

Eventually the drought reached Mesa Verde as well. The major response of the Anasazi was to migrate yet again to new locations. Their descendants are today the Pueblo and Zuni tribes. They still raise maize, squash, and beans, and use irrigation. In fact, when William Smythe started the first large-scale white irrigation projects in the Southwest in 1891, he got the idea because he saw the Pueblos practicing it.[15] Like the Mississippian culture described in chapter 1, the Anasazi apparently began to live less like an empire and more like a loose confederation of villages.

METROPOLIS BEFORE PHOENIX

An even larger urban center existed about a century later than the Anasazi civilization, in central Arizona. It was found in almost the exact location as modern Phoenix.

Once again, we do not know what the people called themselves. Their presumed descendants, including the Akimel O'odham (Pima) and Tohono O'odham (Papago) tribes, refer to them as the Hohokam—the disappeared ones. The Hohokam urban center was the Phoenix of its day, and for the same reason: irrigated agriculture.

The major difference between Anasazi and Hohokam irrigation is that the latter was based on more or less permanent water supplies: the Salt and Gila Rivers. While these rivers can dry up, they are more reliable than the desert washes that supplied water to Chaco Canyon. The Hohokam canal system was extensive. The longest canal was at least twenty miles in length and up to twenty feet deep. Archaeologists estimate that this canal system could irrigate up to 110,000 acres and support a peak population, circa 1300 CE, of between twenty-five and fifty thousand people. These were apparently not religious pilgrims, but permanent residents. They lived off of the same kinds of crops that the Anasazi raised: maize, beans, and squash, origi-

The Blessings of Water 111

nally from Mexico. They also raised a lot of cotton, for textile use, along with agaves, on unirrigated land. Agave stems can be roasted to produce a sugary food. The canals and rivers, permanently wet even when not filled with water, provided another benefit to the people, which the Anasazi did not have: fish for protein.[16]

One of the main reasons that the Hohokam civilization is less well known than the Anasazi is that few of their buildings remain. While some stone walls can still be seen, much of the construction was adobe, which was perfectly suitable for people at the time but which has largely eroded away.

A few of the Hohokam canals can still be seen, but most of them have disappeared. This is because modern concrete-lined canals, which provide water to the thriving Phoenix metropolis, have replaced them— that is, the ones that have not been filled in with dirt. The Hohokam built canals in the most logical places to supply and to use water; they also provided just the right slope to deliver the water with minimal flooding and erosion. Those were the same places where modern engineers have built the Phoenix canal system.

Constructing such a large irrigation system using primitive tools such as wood shovels and without livestock power required an immense amount of social organization. Oppression was clearly involved. I would certainly not have wanted to spend my life digging canals with wooden shovels outside in a desert that today routinely reaches 110 degrees Fahrenheit in the summer. While there is no clear evidence that the whole Phoenix basin was organized into slavery, important canal junctions were dominated by platform mounds, which may have been residences or offices of local administrative leaders. It is clear there was a rich upper class that could import luxury items (including nose plugs, pendants, earrings, bracelets, necklaces, and sophisticated shell inlays) from Mexico, and there were large public ball courts resembling those found in Mexico. There were also decorated shells from the Gulf of California which, unlike the other items, were traded as far away as Cahokia. There is little evidence, however, that these rich rulers had as strong of a religious component to their power as did the rulers of the kiva-ridden Anasazi cities or of the Mississippian civilization.[17]

Starting about 1350 CE, the Hohokam civilization went into decline. Unlike the Anasazi, the Hohokam left little evidence of violence and warfare. The main factor may have been overpopulation. A peak population of fifty thousand was probably far beyond the ability of their agriculture to sustain. The canals themselves, when first built, must have worked beautifully. But occasional floods, as well as prolonged droughts, would have damaged portions of the canal system, and individual canals cannot work wonders

112   *Forgotten Landscapes*

if the system itself malfunctions.[18] While soil erosion might have partially filled the canals, research has shown that irrigation can improve soil quality: The very same erosion that filled the canals might produce superior soil in the former canal beds.[19] At some point, however, maintenance outweighs benefits. The Hohokam migrated away. Their descendants, such as the Akimel and Tohono O'odham, remember little about them.

SIN AND SALINIZATION IN ANCIENT MESOPOTAMIA

We can learn a lot about irrigated agriculture in western North America, and the societies that depended on it, by comparing it with irrigation in Old World history.

In the Old World, agriculture itself got started out in arid lands (chapter 4). Regions well-watered by rain remained the home of hunters and gatherers for many centuries after arid land peoples had developed irrigation. For instance, in China, the earliest irrigation was in river valleys with a dry climate. In Mesopotamia, the earliest agriculture developed in the usually dry floodplains of the Tigris and Euphrates Rivers. Later, irrigation was developed along the Nile in Egypt also. A period of evolution, in which the inhabitants (perhaps unconsciously) selected what would become crop cultivars, preceded the "invention" of agriculture.[20]

In Mesopotamia ("the land between the rivers"), the land was very rich and frequently well-watered because of floods coming from the Zagros Mountains. As the land was not thickly forested, it was relatively easy for early farmers to establish square fields to which water could be conducted by canals, from the Euphrates, through the fields, and draining into the Tigris, much of which is at a lower elevation.[21] Not only was it excellent for raising wheat, lentils, and other crops, but the weather was consistently warm, perfect for orchards. This was the traditional location of the Garden of Eden.

All that was necessary to unlock the food production capacity of Mesopotamia was a massive network of canals and massive government bureaucracies to create and sustain it. Hundreds of thousands of laborers were conscripted to build it, and competition was fierce to get the water. Getting water was a major impetus for wars, and destroying canals was a major military tactic. There was interminable conflict between city-states downstream and city-states upstream, who were accused of taking more than their fair share of the water, particularly in years with low water levels. Also, the system was not designed to last forever. As also happened centuries later in Native America, the Mesopotamian canals filled in with silt, and the channels were damaged by floods. Repairs, in response to a single flood, were expensive and could take years.

Only wise oversight could allow a city-state to persist through inter-ruptions in water supply, in particular saving up grain during bountiful years to use during famine years that might or might not come. According to oral accounts, the Israelite slave Joseph rescued Egypt from famine be-cause he was able to interpret dreams. The pharaoh had a dream in which God revealed there would be a time of abundance followed by years of drought. Joseph, in prison at the time, told the pharaoh to save up grain during the good years to prepare for the coming hard times.

In addition to food shortages during drought years, dead animal bodies could pile up, along with wastes normally washed downstream, leading to outbreaks of disease.

Food production was not the only consequence of irrigated agricul-ture in Mesopotamia. So extensive were the irrigated lands and so great was the production that the officials (probably priests) in charge of irrigation had to keep track of agricultural production and water use. Taxes were paid to the priests on the basis of the value of field productivity. It is almost impossible to know the production of each city, town, or field. Agriculture was undoubtedly an important contributor to the development of writing (at first on clay tablets) and the predecessors of numerals. The Anasazi or Hohokam peoples apparently never ran into this problem.

One of the main problems that led to the eventual failure of Mesopo-tamian irrigation was an invisible one: salinization, or the buildup of salt in the soil. Salt is toxic to most crops. Wheat is very sensitive to salt, barley a little bit less so. The salt buildup occurred partly because of evaporation of irrigation water from the fields, a process that leaves salt molecules behind in the sunlight, and partly because irrigation allows surface water to make a physical connection with salty subsurface water, thus drawing the salt water toward the surface.

Progressive soil salinization is suggested by the historical record. The earliest Mesopotamian civilization was Sumeria, furthest downstream, near the Persian Gulf, while the later Mesopotamian civilizations were farther northwest and upstream, such as Babylon (Babylonia) and Nineveh (As-syria). This suggests a progressive abandonment of salinized soil. No one could see this happening—unless the salt crystals dried out on the surface—but even then, no one thought of keeping a record of it. We would never have known about salinization in Mesopotamia were it not for archaeolo-gists such as Thorkild Jacobsen.[22]

114   *Forgotten Landscapes*

The indirect technique that Jacobsen used to estimate how much salt was in the Mesopotamian soils was to quantify the amount of grain brought to local temples as taxes. In particular, he was interested in the ratio of salt-tolerant barley to salt-sensitive wheat. Wheat was the preferred crop. A barley-to-wheat ratio of zero meant there was essentially no salt in the soil. As salt built up in the soil in a few fields, temple records would show a low barley-to-wheat ratio of offerings. As salt built up in more and more fields, the ratio would increase. In any given location, most or all of the offerings could be barley. This was due to salinization. It undermined food production and weakened the Mesopotamian economies, ultimately contributing to their collapse.

This was a problem that did not occur in Egypt. The primary reason was that, despite Egyptian irrigation, salty soil never built up. Egyptian astronomers could predict the coming of Nile floods each year, which would bring water and soil from the Nubian mountains upstream. Farmers would allow the floodwaters to flow over their fields for a few weeks. This would deposit a new layer of soil. Any salty soil from the previous year's irrigation was washed away or buried.

It was also not a problem that occurred very often, as far as we know, in ancient Israel. Israel was hilly country, so the water drained downhill. Soil erosion was partly prevented by building retaining walls, which created crude terraces. The Israelites had to do this because, despite the paeans of praise that the Old Testament chroniclers piled upon Israel's supposed military triumph when they left Egypt, the surrounding nations, such as the Philistines, occupied the productive lowlands. The retaining walls had to be carefully maintained because once a breach began, erosion quickly and disastrously widened it.

The Israelites learned the techniques of terracing, and of building cisterns, from the people who were already there and whom they supposedly slaughtered. They even learned from the "Canaanites" about a practice in which they let their fields lie fallow every seven years ("Sabbath of the fields"; see chapter 4). Although there was not much salinization in ancient Israel, their efforts to prevent soil erosion proved wanting, especially during times of war and social instability. Even, said the prophet Joel, the beasts of the fields cried out to the Lord because the creeks dried up due to a combination of natural factors and human mismanagement.[23]

The last Mesopotamian empire, the Babylonians, eventually fell due to war and unsustainable agriculture. The Israelites, whose Old Testament writings remain some of the most influential literature in the world today,

The Blessings of Water   115

interpreted this as God's judgment upon them for their pagan sins. But it wasn't just sins. It was also salinization.

One of Israel's neighboring tribes, the Nabataeans, most famously the city of Petra, lived in the desert just to the south. Petra is in modern Jordan. Impressive buildings were nestled in caverns in narrow canyons. To walk through Petra today is a breathtaking experience, and it must have been even more so before the collapse of Nabataean society.

But the stone walls were not the most amazing thing about Petra. Like the Native Americans, the Nabataeans had to manage their water resources carefully. They built lots of catchment areas to save water for irrigation, and a complex system of cisterns for municipal water. Enough water accumulated in a lagoon near the top of the mountains that water flowed into the city under pressure, creating fountains. The Nabataean civilization collapsed not due to erosion or salinization but due to conquest and destruction by enemies.[24]

Salinization does not appear to have occurred in Anasazi or Hohokam fields. Perhaps this was because the soil was more porous, or because irrigation was less intense, in America than in the Old World, or that it did not continue long enough. What we can safely assume is that in America, as in Petra, the ultimate collapse of the system was not because there was anything wrong with irrigation, but because of war and social disruption. The disappearance of Anasazi and Hohokam power was not through any lack of technical skill among the Native Americans.

WATER AND POWER

In the American Southwest today, water is the single most important fact of life. It overwhelms every other factor. The entire future of this area depends on solving the issues of access to water. The stakeholders in this conflict are not only several states, and their municipal economic powerhouses, but also governments at every level, and several major Native tribes.

This conflict would seem to be an excellent opportunity to increase water-use efficiency. Around the world, many municipalities have found ways to do more with less water.[25] The old style of irrigation was to pour water onto thirsty land. But most of this water evaporated into the air, wasted, and leaving a salty residue. This was the kind of irrigation I saw, and that my family used, while I grew up in the San Joaquin Valley (chapter 8). But Israeli scientists developed a method of drip irrigation,

116   *Forgotten Landscapes*

which delivers water only and precisely to the places it is needed, at the base of each plant, as much as possible. Drip irrigation not only reduces the amount of water needed to produce each dollar's worth of agricultural output, but it also reduces salinization.

Water-use efficiency can be enforced in building codes for new construction, but is difficult to incorporate into old buildings. In addition, a lot of water-use efficiency requires voluntary, society-minded cooperation. Water-use efficiency in irrigation can be indirectly enforced by charging higher fees for the water from irrigation systems. But, while many thousands of people are willing to increase their own water-use efficiency, at least as many people are openly hostile to reducing their water use for the community good. And in California, large farmers (each of whom can control many thousands of acres of land) have almost unassailable political power, and most of them want to use water, not conserve it.

The Colorado River

The Colorado River flows through and between seven states, making its control a federal matter. But it is up to each state to come up with ways to meet the federal requirements. And what happens if a state refuses to comply? What sort of force, if any, should the federal government use in response?

The seven states reached a historic agreement, called the Colorado River Compact, in 1922. But by the end of the twentieth century, economic and natural situations had changed. Perhaps most importantly, the amount of water flowing in the Colorado River was significantly less than it was in 1922. This is most clearly visible in the photographs that show Lake Mead, a major reservoir of Colorado River water, rimmed with lime from falling water levels. Also, by the end of the century, Native tribes were finally allowed to assert a limited amount of power for their own interests, rather than having those interests entirely represented by the state governments or the Bureau of Indian Affairs.

In a major 2023 breakthrough, the seven states agreed to a redistribution of water, in which California, Arizona, and Nevada approved a 13 percent reduction in the use of Colorado River water.[26] Major problems remain—in particular, how to prepare for the scientifically inevitable effects of global climate change.

Although the descendants of the Anasazi and Hohokam irrigators are considered to be minor players in the water wars of the Southwest today, other tribes such as the Navajo have large reservations and populations.

The Blessings of Water   117

California Water Wars

Although the Colorado River is the only source of water for much of the Southwest, the same is not true of California, in which several large rivers carry snowmelt down into the Central Valley. As I will describe in chapter 8, there is so much water that the Valley floor used to be covered with wetlands. The snowmelt in the Central Valley comes from the Pacific slope of the Sierra Nevada, which used to have, and sometimes still has, massive snowpack in the winter. There is also a little bit of snowpack on the eastern slope, which drains into the Owens River. Despite this seeming overabundance, water has been the source of intense political and economic conflict throughout California history.[27]

The Central Valley water is used mostly for agriculture, although a few large cities such as San Francisco and Sacramento use it for municipal supplies. But even these water supplies led to intense conflict. The city of San Francisco wanted to build dams in the Sierra Nevada to flood mountain valleys as reservoirs. Their major opponent was one of the first environmentalists in America, John Muir, who fought to oppose dam construction. He succeeded in saving Yosemite Valley, partly because he invited President Theodore Roosevelt to see how beautiful it was. But he did not succeed in saving a lower valley, Hetch Hetchy, which is today a reservoir, and the major source of water for San Francisco.

But the part of the state that has the greatest craving for water is Los Angeles and surrounding cities. For a century, water availability has been seen as the greatest limitation on the economic growth of the city. The low mountains around Los Angeles are dry and usually have no snowpack.

William Mulholland, in charge of Los Angeles water during the 1920s, found what he thought was the perfect solution.[28] To the northeast, snowmelt from the eastern Sierras drained into Owens Valley, where few people, mostly farmers and ranchers, lived. He wanted to get the Owens Valley water for Los Angeles. But to do so, he had to convince the landowners in the Valley to give up their rights to the river water. In what today is considered a towering example of political intrigue, Mulholland managed to do just this.

Nevertheless, California water policy remained a patchwork of selfish enterprises, of which the Los Angeles water grab was only one. Once the Great Depression began, President Franklin Roosevelt started the New Deal. One of its major purposes was to put unemployed people to work. One way of doing this was to revamp the entire irrigation system of the western United States. One of the projects that resulted from this, a decade after Mulholland, was the Central Valley Project of the Department of the Interior. A series of interconnected canals supplied water in an organized fashion to the entire Central Valley. It is difficult to overestimate the sense of power and pride that was represented by the large irrigation projects of the early twentieth century in general, and the Central Valley Project in particular. Another New Deal agency, the Federal Writers' Project, hired unemployed writers to produce a book that sang the praises of this irrigation project.

But even the story of federal involvement in California irrigation was complicated. Two federal agencies were involved in building dams in the lower Sierra Nevada mountains, and both of them distributed water from the resulting reservoirs. The main charge of the Army Corps of Engineers was to prevent flooding, especially in the towns occupying what used to be the bed of the vast lakes that covered the Central Valley. The Bureau of Reclamation (of the Department of the Interior), the builder of the Central Valley Project, had the more direct purpose of providing irrigation water for agriculture and municipalities. They were in a race to see who could build the most dams and canals.

And the state of California also got involved. The California Aqueduct took water from Northern California and the Central Valley into the Los Angeles area. This was an enormously expensive undertaking which required pumping millions of gallons uphill from the San Joaquin Valley over Tehachapi Pass (figure 5.7). As Marc Reisner said, in California, water flows uphill, toward money.

Figure 5.7. Water from the California Aqueduct is pumped uphill over the Tehachapi Mountains in California. Looking southward from the south end of the San Joaquin Valley. *Author photo.*

PRE-CONTACT NATIVE AMERICANS AND WATER

Native American tribes did not transform every arid region in which they lived into agricultural landscapes through directed use of water or irrigation. Sometimes they just left it as it was.

As I will describe further in chapter 8, the Central Valley of California had vast expanses of wetlands teeming with fish and wildlife. It may have seemed like a wonderland, but not very many Natives actually dwelt near the wetlands. It was not easy to hunt elk there, as the wetlands were teeming with mosquitoes, and the dominant plants—sedges and tules—were not edible. Most of the Native inhabitants lived in the oak woodlands of the foothills, existing primarily off of acorns and fruits, as discussed in chapter 6.

The soil in the Valley bottom was deep and rich, but was (except briefly in the winter) dry and hard. The Natives did not know there was plenty of water available for irrigation, sometimes not far below the surface. Even if they had known it, they might not have considered it worth the effort to dig canals in order to irrigate . . . to irrigate what? Unlike the Anasazi and Hohokam, they did not know about maize, beans, and squash from Mexico. They might have irrigated pigweed like the Paiute, but it would hardly have been worth it, with a bounty of wild grains and tubers from the fire-managed grasslands (chapter 2) as well as acorns available to them.

Ironically, as a plant ecologist who worked in Owens Valley explained to me, it might have been "dirty politics as usual," by which Los Angeles stole the Owens Valley water, which protected the natural habitat of Owens Valley from exploitation. Had it not been for William Mulholland, Owens Valley might have been another Los Angeles, with hotels, theme parks, and extensive tracts of suburbia. Might there have been a water slide down the east slope of Mount Whitney? Instead, you can look upslope nine thousand feet from the Owens Valley floor to the tops of the tallest North American mountains south of Alaska (figure 5.1) and see mostly natural vegetation.

Now, as then, brine shrimp in the alkaline waters of Mono Lake feed flocks of possibly a million eared grebes at peak migration. The major difference is that the water level of the lake is declining, leaving behind mounds of alkaline crystals. According to the signage at the lake, this resulted from water export to Los Angeles, which would have lowered the water table. Aside from this, Owens Valley doesn't look all that different today from back when the Paiutes irrigated fields of native grasses and pigweeds, ate fat caterpillars of the pandora moth from the pine trees, and feasted on the abundant larvae of brine flies that filled Mono Lake.[29]

6

LITTLE PARADISES

Orchards in Prehistoric Native America

I stood in a bottomland hardwood forest in Oklahoma near the important Native American site of Spiro Mounds (chapter 1). All around me, the largest trees were pecans (figure 6.1).

The trees were large, but none of them were more than a few decades old. Although they started growing long after Elizabeth arrived in Oklahoma on the Cherokee Trail of Tears (see the introduction), the grove itself may have already been here. In fact, this grove may have been present when the Mississippian culture built Spiro Mounds, and the Natives who came to Spiro might have eaten nuts from this very grove. Pecan trees produce large crops of delicious nuts and have been bred for cultivation since the late nineteenth century. Today, American pecans are famous all over the world.

Native Americans loved pecans. Nuts are one of the best sources of oils and protein, aside from animal meat. Pecans have up to 73 percent oil content and taste as good as the walnuts, hickories, and chestnuts—and better than the acorns—that the Natives also ate. They gathered pecans from natural trees. They still make a ground paste known as *kanuchi* from the nut meats, pounding the nuts (after removing the husks) in hot water in a basin created from a hollowed-out tree stump. The oily nut meats rise to the top of the water where they can be skimmed, dried, and stored. Kanuchi resists spoiling the same way butter does; it is mostly oil and protein, with very little water, which bacteria need in order to spoil food. Like butter, the kanuchi could go rancid upon prolonged exposure to the air, but it stayed fresh long enough to store for the winter or to be taken on a long journey on one of the many trade routes that bound the Mississippian culture together. To make kanuchi soup, just stir a kanuchi ball into hot water.

The traveling Natives were happy when they found healthy, productive natural pecan groves. It is very likely they would also have carried

Figure 6.1. A bottomland hardwood forest in Oklahoma, at the site of Spiro Mounds. The dominant trees are pecans (*Carya illinoiensis*), some of which may be the direct descendants of pecans planted by the mound builders of the Mississippian Culture; with an understory of boxelder (*Acer negundo*). Has this always been a natural forest, or was it once an orchard? *Author photo.*

Little Paradises   123

intact pecans with them to plant under appropriate conditions (in moist soil along rivers and creeks) in new locations so that they could have even more pecans. As a result, by Mississippian days, and certainly prior to European contact, some of the pecan groves would have been natural, while others were orchards planted by the Natives. Perhaps most of them were natural groves that had been expanded by Natives planting pecan seeds nearby—that is, groves that were both natural and artificial. I looked at "wild" pecan groves a different way once I realized they resulted from the activities of both nature and humankind.

A Native orchard would not have looked at all like what we would consider an orchard today. Modern pecan orchards have trees all the same size, planted and spaced in straight lines, and with no undergrowth. In contrast, a Native American pecan orchard would have looked, to our eyes, like a wild grove. It would have been a natural forest that the Natives nudged in the direction of yielding more of a useful product, in this case pecans.[1]

As I explain in other chapters, Europeans considered themselves superior to Native Americans because, they thought, the Natives had no agriculture. Gardens, but not agriculture. And no orchards either. The European error was that they looked right at Native crop fields and orchards and did not see them. If you define an orchard as a place where woody plants have been either planted or manipulated to yield useful products, then orchards were all over the place in pre-Columbian North America.

Just as they did with Native crops (chapter 4), Europeans destroyed Native groves as part of their conquest of the supposed savages. As William Apess, a Native American, said in 1836, King Philip predicted that white conquerors would "not only cut down their groves but would enslave them." King Philip was a Wampanoag chief who led a short-lived attack on English colonists in 1675. Were these "groves" to which Apess referred just the forests that the Wampanoag loved and the English feared, or were they orchards?[2]

MODERN ORCHARDS

Orchards bear just enough of a resemblance to forests that we can feel close to nature in them. This has been true for thousands of years. The Garden of Eden, the walled Persian gardens called pairidaezas (paradise), and Kubla Khan's Xanadu were orchards with various tree species useful to humans planted together and with streams of water flowing through their shade. I remember, as a child, trespassing through orchards in the

124   *Forgotten Landscapes*

San Joaquin Valley. I thought the orchards of that artificial paradise were incredibly beautiful and peaceful (chapter 8). While they could not compare to the giant sequoia trees in the nearby mountains, they were a reasonable substitute for a little boy to visit while riding his bicycle.

Aside from this, however, orchards of our modern industrial food production systems are not like natural forests at all.

- The fruited plain of the San Joaquin Valley contains patchworks of orchards, each of which, unlike Eden and paradise and Xanadu, is a single species: separate orchards of oranges, other citrus fruits, olives, almonds, cherries, walnuts, and even Japanese persimmons. A forest, in contrast, consists of a mixture of tree species.
- A modern orchard has virtually no undergrowth, while a forest has numerous understory species of woody and herbaceous plants. To suppress the growth of "weeds," many herbicides are used in modern orchards.
- The most productive modern orchards consist of single tree species set down in straight lines, all planted at the same time and therefore of the same size. The trees are spaced just right to allow the maximum growth without "wasted" space. In contrast, most forests have trees growing wherever the seeds happen to land and of all different sizes and ages.
- Frequently, the modern orchard trees themselves are individually altered. Many of the orange trees in the San Joaquin Valley are seedless navel oranges. They are propagated by cuttings from existing navel orange trees grafted onto a rootstock of a hardier orange species that produces fruit too bitter to eat. English walnuts cannot grow well in the soils of the Valley. Cuttings of English walnuts are grafted onto rootstocks of native black walnuts (figure 6.2).
- In modern orchards, as with other kinds of monoculture, a single species invites pest damage, which is usually controlled by the massive use of pesticides.
- Not only have most natural species been eliminated from the modern orchards themselves, but from around them. Modern orchards are often set down in a bare environment.
- Modern orchard trees have been bred for synchrony in fruit production, so that all the fruit ripen at the same time, for ease of commercial harvest. In Native orchards, each tree produced its fruits or nuts at slightly different times, and the harvesters had to visit them repeatedly.

Figure 6.2. This orchard of English walnuts (*Juglans regia*) consists of trees whose branches were grafted onto the rootstocks of native California black walnuts (*J. californica*). The trunks of the two species grow at different rates. *Author photo.*

126   *Forgotten Landscapes*

Because a modern orchard contains only one species, and may be isolated from most other plant species, the pollinators cannot find enough nectar and pollen to consume except during the brief time that the orchard is in flower. While some commercial tree species can produce fruits and nuts without pollination, and while a few of them are pollinated by the wind, most of them require insect pollinators. The most important example in modern American agriculture is almonds. While an almond orchard in bloom, buzzing with bees, is an inspiring experience, it only lasts a couple of weeks. What will the bees do the rest of the time?

In a forest, the pollinators will find a different species of flower to pollinate. They shift from one species to another throughout the growing season. For example, in the wild fields I have studied in Oklahoma, insects pollinate smooth sumac bushes (*Rhus glabra*) in late May, but by late June they are pollinating winged sumac bushes (*Rhus copallinum*). Each flower species has its own blooming season and does not have to compete with other flower species for pollinator services.

When an almond orchard needs bees, it needs them in large numbers, and it needs them now. In an industrial society, an ecological problem is often addressed by a technological solution. In this case, beekeepers truck in whole colonies of honeybees (each in their own white wooden box) and leave them in the almond orchard during flowering. Then they haul them to some other orchard, perhaps an orange orchard, that flowers at a different time. Between orchard flowerings, the beekeepers find large fields of sweet clover (genus *Melilotus*). Bees store honey as food, but they cannot just stop foraging, sit around, and take life easy, except in winter.

Other crops, not just orchard fruits and nuts, need pollinators. Every third bite of food from our agricultural system comes directly or indirectly from crops that require pollinators.[3] For example, humans do not usually eat alfalfa and clover, which require bee pollination, but they are important for raising the cattle that produce beef and milk.

This technological solution—trucking in a large number of hives— creates new opportunities for the system to break down. One example is that the high population density of bees encourages the spread of parasitic mites that can kill the bees, and that do not spread very much in wild bee colonies.[4] Also, because bees provide such a valuable economic service, growers compete for the opportunity to hire the services of beekeepers. But when the keepers leave the hives in a field or orchard overnight, "bee rustlers" may steal the hives.[5]

I think it is very safe to say that bee mites and bee rustling were not a problem that Native Americans faced with their orchards.

NATIVE AMERICAN ORCHARDS

Most modern people are probably unaware that Native Americans even had orchards before European contact. You may have heard that they had orchards after European contact—for example, the European peach orchards raised by the Navajo on the thin beaches beneath canyon walls of the Colorado River and its tributaries. Destruction of the peach orchards was an essential part of Kit Carson's conquest of the Navajo in 1864.[6] The Cherokees also had European peach orchards which they left behind when they were forced onto the Trail of Tears. These orchards were not what Native orchards were like prior to European contact.

The orchards that the Natives had before European contact differed in almost every way from modern ones. They would have looked more like forests. They had undergrowth, such as the box elder trees underneath the pecans at Spiro. Native orchards might have had all the same species as natural forests, but with different relative abundances—for example, pecan orchards might have been just like pecan forests but with relatively more pecan trees. We will never know, since we have no truly natural forests to which they can be compared.

In most of the examples below (and summarized in table 6.1 at the end of the chapter), the plants can also spread by aboveground or underground stems once even a small population has become established by seed. What was true for trees would also be true for shrubs and vines with edible fruits.

EDIBLE FRUITS

Mulberries, Blackberries, and Grapes

The seeds of most species of woody plants with juicy, delicious fruits will not germinate unless they have passed through an animal's digestive tract, where the digestive juices soften the hard seed coats. Only then can the seeds germinate. This is also the evolutionary reason that fruit pulp (not just of prunes) usually has laxative properties. The laxatives help to make sure that the seeds pass through the animal's digestive system quickly and do not remain there long enough to die in the anaerobic and toxic conditions.

Animal seed dispersal via intestines is good enough for non-human animals to spread the seeds to new, though not distant, locations. The fruits that humans like to eat are often perishable and only briefly available. Natives would often dry the fruits and carry them for long distances, either

128 *Forgotten Landscapes*

alone or as a component of pemmican. When a human eats the dried fruit, the seeds come out in the waste in a distant location where, perhaps, the species had not previously grown. This is probably what happened with small-seeded fruits, such as mulberries, blackberries, and grapes. For these juicy fruits, humans and non-human animals were little different in their effect on the abundance and range of the plants, except that humans traveled much farther. Many edible fruits might have had their abundance enhanced by germinating in latrine areas outside of Native villages.

The first white visitors to what are now the Southern states remarked on the abundance of mulberries (genus *Morus*). When Hernando de Soto visited a tribe that may have been Cherokee around 1540, they presented him with a basket of mulberries as a gift. These would have been the native species of red mulberries, rather than the white or black mulberries, which are naturalized imports from Asia. Mulberry fruits are famously delicious. How much of this abundance of mulberries was due to birds, and how much to humans, cannot now be known. Captain John Smith in Virginia wrote in 1625: "By the dwelling of the savages are some great mulberry trees, and in some parts of the country, they are found growing naturally in pretty groves."[7] Smith's dismissal of the Natives as savages blinded him, but only partially, to the extent they might have planted trees that were very useful to them.

Pawpaws and Persimmons

Some of the favorite fruits of the Natives had large seeds, which they would not have swallowed. Consider the native pawpaw (*Asimina triloba*), which produces a delicious fruit that reminds some people of starfruit or banana. Pawpaws were rare except in locally abundant patches—patches that might have started growing near Native encampments.[8] In Tulsa, Oklahoma, the Muskogee tribe has planted a garden around the Council Oak, a post oak tree where their tribal leaders met in 1836 after arrival on their Trail of Tears migration, two years earlier than the Cherokee Trail of Tears. The garden features medicinal and food plant species important to the tribe. I saw several pawpaw trees maintained in this garden. Natives also thought the fruits of persimmon (*Diospyros virginiana*) were delicious. Persimmon seeds are also pretty large.

The most important persimmon consumers are coyotes and raccoons.[9] These animals will swallow the entire fruit, thus allowing the seeds to soften in their guts. I have seen lots of raccoon droppings filled with persimmon seeds that, unlike seeds taken directly from the fruits, germinate readily.

Little Paradises 129

Raccoons have nimble fingers and could possibly remove the seeds before swallowing the fruit, but the pulp sticks very tightly to the seeds, and raccoons seldom have time to be finicky eaters. Humans, however, are different; we spit out the seeds without giving their coats a chance to soften in our guts.

Plums also have seeds that humans would be unlikely to swallow, but the Iroquois planted plum trees near their villages.[10] Small fruits like plums, but not large ones like persimmons, may have been eaten by passenger pigeons, and would as a result have been abundant even without human help. The pigeons were always abundant, even though as chapter 3 explains, the huge pigeon populations of the eighteenth and nineteenth centuries probably resulted from the crash of Native American populations due to European diseases, who had previously hunted the pigeons and kept their populations small.

How, then, might the Natives have dispersed persimmon and pawpaw seeds to new locations? There are at least two possibilities. First, it is possible to abrade the seed coats (a process known as *scarification*) to get them to germinate without passing through an animal's intestines. The abrasions allow water to get into the inside of the seed, where the baby plant lies dormant. The Natives—especially children playing with seeds—might have figured out how to scarify the seeds before planting them in a new location. Second, raccoons and coyotes do not normally travel great distances in their home ranges, but they might have chosen to run along the network of trails, and in so doing, traveled farther. In addition, raccoons would have found persimmon and pawpaw seeds with some of the pulp still on them in human midden piles, eaten them, and spread them. In this way, the Natives might have indirectly promoted the dispersal of pawpaws, persimmons, and plums to new locations inside the guts of raccoons and coyotes.[11]

Dates

Perhaps the most surprising example of a possible prehistoric Native American orchard is the New World version of a date grove.

Date orchards were some of the earliest examples of Middle Eastern agriculture. They figure prominently in Egyptian hieroglyphics. The fruits of the date palm are sweet and can be made into delicious bars that do not readily spoil. You can buy date bars in the supermarket that are not very different from what the Egyptians might have eaten. Dates and honey were about the only sources of sweetness in the ancient Middle East.

Old World date palms are now grown in the desert regions of California, especially the Imperial Valley. We know why they are there. They were planted as a commercial source of dates after irrigation began. But just a few miles west of the Imperial Valley, in the Anza-Borrego hills, you can find native date palms—the California fan palm, *Washingtonia filifera*, an entirely separate species with fruits surprisingly similar to Old World dates.

We scientists and nature lovers thought we knew why the native fan palms were there. The way I heard the story when I was studying California plants as an undergraduate was that palm trees were widespread in the desert regions of eastern California during earlier eras of greater rainfall. Then, as the rainfall decreased, most of eastern California became a desert. The date palms retreated to desert springs, where they had enough water to survive. Besides desert springs, there are also arroyos, where rainwater flows, sometimes rapidly, during the brief desert rainy season. The arroyos themselves dry out, but the soil underneath them remains moist for a longer period than the desert soils around them. Moisture is the reason you find little groves of fan palms at desert springs, the story goes. If you take the trail at Anza-Borrego Desert State Park, you can visit one such grove (figure 6.3).

But some of the facts do not fit this interpretation. The fan palm is almost the only species of tree in these groves. One might expect that if

Figure 6.3. A grove of *Washingtonia filifera* palm trees cluster at a desert spring in the Anza-Borrego hills east of San Diego. *Author photo.*

Little Paradises   131

these groves were remnants of previously widespread trees, there might be a few other species there also. There are certainly lots of other woody plant species that grow in California desert oases and arroyos. Examples include the following:

- Palo verde and various species of desert acacias. Acacias often have green bark (*palo verde* means "green stick" in Spanish), which gives them additional photosynthesis when they drop their leaves during the dry season. The giant saguaro cactuses start their lives growing in the shade of palo verde trees, then outgrow them, towering into the air long after the acacias have died. The acacia is an example of what is called a "nurse plant" since it helps the young cacti to grow.
- The desert willow. Willows do not have conspicuous flowers, but this tree is not really a willow. A relative of catalpa trees, it produces striking magenta flowers and is a popular horticultural plant in dry areas, so long as its roots can reach water. If you see a tree that looks like a willow in a parking lot in the western part of the country, it is probably a desert willow.
- The smoke tree is so called because its fruits have long, curly extensions on them that look like smoke rising from the branches.
- The elephant tree is a succulent tree, so called because its bark reminded somebody of pachyderm skin.

These trees are all smaller than fully grown fan palms, which would have outgrown them. But a few of these other woody species would undoubtedly be mixed in with the palms if the palm groves were remnants of natural woodlands that preceded the deserts. The others grow in desert washes all around the fan palm grove. Why are they largely absent from the palm grove? And where are the palm trees that you would expect to find in the arroyos?

James W. Cornett suggests that the reason is not entirely natural.[12] He points out that there is no geological record of these palm trees in the area from before the Natives arrived. The most likely explanation, says Cornett, is that Natives planted the trees, then maintained the grove by fire. These palms are moderately resistant to fire; in figure 6.3 you can see one of the trees that has partially burned but is still alive—something I did not notice when I originally took the photo. The Natives of the Mojave, Pima, Paiute, and Cahuilla tribes started and maintained this and other similar groves in the desert for the same reason modern people raise date palms: The fruits are delicious and nutritious.

132   *Forgotten Landscapes*

The fan palms were not just useful for their fruits. The Natives used the tough leaves to make sandals, roof thatch, and baskets. The leaf bases are big and strong and have serrations that can be used like knives.

Once started, the palm orchard would maintain itself by fires, even after Natives quit burning them deliberately. It is almost impossible to keep the dead leaves which accumulate next to the trunk from catching fire.

Native orchards, like this one, might be hiding in plain sight.

EDIBLE NUTS

Chestnuts

About two thousand years ago, suitable conditions for American chestnuts (*Castanea dentata*) existed in only about 4 percent of the oak–hickory forests of eastern North America. By the time of European contact, chestnuts covered 40 percent of the land. What happened to make the chestnuts become more common? Forest ecologist Thomas M. Bonnicksen suggested that Natives gathered chestnut seeds to eat, but also threw many of them onto suitable ground to produce chestnut orchards.[13]

We seldom think of American chestnuts today, because a fungus accidentally brought in from Europe killed almost all of them along the Appalachian Mountains, starting in the North around 1900 and ending in Georgia around 1950. All you can see today is small shrub–like resprouts from chestnut root stock. But at one time chestnuts were an important component of the drier part of Eastern forests.

Bonnicksen also suggested that the controlled use of fire by Natives (chapter 2) created more open, dry conditions where chestnuts can out–compete maples and beeches.[14] Native Americans probably threw chestnut seeds on the ground after burning an area. The Natives loved chestnuts, perhaps even more than Europeans love their own species of roasted Christmas chestnuts. The Natives just had to pick the nuts up off the ground, although they had to race against the squirrels to get them. If you eat a nut, you destroy the seedling. But the nuts will germinate in the ground. Squirrels hide nuts in places where they hope they will be able to find them later, and some of them get lost and sprout into new trees. Humans can spread uneaten chestnut seeds, too.

Pine Nuts

Far away from the Mississippian culture, Native Americans in the Intermountain West depended heavily on nuts from piñon pines. Before they

Little Paradises 133

cultivated corn (which was impossible in some of the higher and drier areas), they foraged on pine nuts. Many communities continued to do so after they adopted agriculture. Sometimes they made cornmeal cakes and put piñon paste on them. They loved pine nuts, which they could gather abundantly by picking them up off the ground or by striking the piñon branches with oak poles, dislodging the seeds from the cones. The seeds were only briefly available in the autumn, and the Natives would store them in bags or cellars as their primary winter food. They had to race against the squirrels and other mammals that loved the piñon seeds as much as did the humans. They had to grind away the husks before eating the seeds. The nuts could have up to 71 percent oil content. The people sometimes roasted them before eating them, which might have removed some of the turpentine-like flavor. They carried them on journeys as an important food source. Seed-gathering time was their principal tribal celebration.[15]

Gathering pine nuts was not a simpleminded task. The cones were covered with pitch, which was quite an inconvenience to the gatherer. An individual tree, or a whole grove, does not produce a large number of seeds every year. After producing a bumper crop of seeds, the trees had to store up their resources for a few years. Many tree species produce a large number of seeds during "mast" years. The animals that eat the seeds would find that, during mast years, there were more seeds than they could possibly eat; then, in subsequent years, the animals (which had just begun having a population explosion from the abundant food) would not find enough to eat. This process, called *satiation*, allowed at least some of the seeds to escape being eaten during mast years. It usually worked, except when the animals were intelligent humans who kept track of which pine groves produced abundant crops and were therefore unlikely to do so again the following year. The Natives also scouted out the groves to see which ones had lots of immature green cones and might therefore produce a big crop the following year. At the annual celebration, the tribes could plan ahead, knowing how abundant, or how scarce, their food supply would be for the following year.

It seems unlikely that the Western Natives would have planted any piñons. They certainly would have known how, but there might not have been any reason for them to do so. Piñons are one of the dominant tree species over a large part of western North America, and people could have found the nuts anywhere they went. What point would there have been in planting them? Evidence indicates that piñons were abundant even when Native populations were still sparse soon after their arrival from Siberia. When they took piñon seeds with them on journeys, the Natives usually roasted them first, thus killing them. In this way, planting piñons, unlike

134   *Forgotten Landscapes*

planting otherwise scarce trees such as pawpaws and persimmons, would have made no sense.

Scientific attention has focused particularly on a small population of piñons found in Owl Canyon, over 150 kilometers north of any other piñon woodlands.[16] How did this isolated population get started? C. W. Wright suggested in 1952 that the seeds may have been dropped or planted by Natives who were carrying them on a trade route. More recent studies indicate that Wright's idea cannot be proved or disproved. The genetic characteristics of the Owl Canyon population suggest that it started in a single founding event (*dispersal*) rather than being a remnant of an earlier, widespread population (*vicariance*). However, the founding population would have needed a lot of seeds in order to account for the genetic variability that is found in the population today. It is unlikely that Natives on a trade route would have dropped, accidentally or deliberately, a large number of piñon seeds. Of course, how else might the seeds have gotten there? We will almost certainly never know. If the Owl Canyon evidence is all we have to go on, we can conclude that Natives seldom if ever enhanced the abundance of piñons by dropping or planting them. Therefore, there were probably no piñon nut orchards.

Acorns

Even farther away from the Mississippian culture, Natives manipulated oak groves in California. Acorns dominated the cuisine of many Native California tribes. There are lots of oak trees in the East, as well, but chestnuts, pecans, hickories, and walnuts are easier to eat, and taste better. Although the riverside forests of the San Joaquin Valley had a native species of walnut, the California foothill Natives had little choice but to eat acorns. Though acorns are difficult to eat raw, Natives had many ways to process them, often by grinding them up and leaching the bitter chemical compounds out of them with hot water, through leaf filters. They chose granite outcrops for this, and you can still see their grinding holes (figure 6.4). Either the acorns or the meal could be stored for later use. Acorn meal, which has lots of protein, is very nutritious. Though the Natives may not have actually planted acorns, they did take care of the groves, even pruning dead branches away and eliminating trees that might compete with the oaks. Tribes and villages protected their favorite oak groves—that is, treated them like orchards.[17]

Figure 6.4. At Hospital Rock in Sequoia National Park in California, you can still see the mortar holes that the Monache Natives used for grinding acorns into flour. They probably poured hot water through leaf filters to leach tannins from the acorn flour.
Author photo.

136   *Forgotten Landscapes*

OTHER USES OF ORCHARD TREES

Native Americans may have spread other tree species for reasons other than eating the fruits and nuts. I give here six possible examples.

Alders for Medicinal Bark

Alders are large bushes, or small trees, that grow widely across the Northern Hemisphere. They are famous in the traditional folk medicine of many cultures because extracts from the bark can reduce infection from wounds. From my own research I have found that chemical compounds from at least one of these species, the seaside alder (*Alnus maritima*), readily kill staph bacteria, one of the major kinds of bacteria that infect wounds, including strains of multi-drug-resistant staph. These compounds, which I have not identified, might be found in other alder species as well.

The seaside alder has a strange geographical pattern. This species consists of three subspecies. One subspecies grows in wetlands around Chesapeake Bay in Delaware, Maryland, and Virginia (a region often called Delmarva). Another grows along the Blue River and nearby creeks in Johnston County in south central Oklahoma. The third grows in a single set of swamps in Bartow County in northwestern Georgia. That's it. No other place in the world. Such a puzzling geographical pattern irresistibly draws the attention of scientists like me. This pattern of rarity is in stark contrast to other alder species, such as the hazel alder and the red alder, which cover many thousands of square miles. How did the strange geographical pattern of *Alnus maritima* get started?

This brings us back to the concepts of vicariance and dispersal. First, vicariance. The three alder populations might be the remnants of a much bigger population in the past. This species might have grown all over North America after the most recent ice age, and then died everywhere except these three places. This is the explanation that fits the DNA analysis.

There are some problems with this explanation. If the three populations are remnants of a once widespread species, why would the species have survived only in those three locations? Those three places are not, in fact, the best places for the alders to grow. The seaside alder can withstand very cold winter temperatures. Why, then, is it absent from all parts of the continent with really cold winter temperatures, and found only in three locations with relatively warm winters?

One dispersal explanation is that the species began in one of the three places and then dispersed, or spread, to other locations. This was the explanation that I toyed with for a while. I thought that the Delmarva population, which is the largest, was the original location, and that Natives carried the seeds with them to the other two locations. Perhaps it was even Cherokees, I speculated. Cherokees originated north of Delmarva before moving south to their pre–Trail of Tears homeland in Georgia, and finally were driven to Oklahoma—exactly the three locations where the seaside alder populations are found today. This explanation has the problem of many scientific hypotheses. The facts shot it down. If this explanation were true, then the DNA in the Georgia and Oklahoma populations would be subsets of the DNA found in Delmarva, which is not the case.[18]

But I am not quite convinced that I should surrender the dispersal explanation. While it is clear that the idea of Delmarva as the source, with mythical Cherokees carrying the alder seeds to Georgia and Oklahoma, is wrong, who is to say that Delmarva had to be the original source? Maybe the original source was some other location where it has since become extinct, from which all three modern populations have come? Perhaps this other location was known centuries ago to Natives but is now unknown to modern scientists (as well as modern Natives). After all, the Georgia population was hiding in plain sight until it was "discovered" by scientists in the 1990s. Only a handful of us botanists would recognize it if we saw it growing in the wild in some other location. Other botanists might assume, if they saw it, that it was a hazel alder or a red alder.

It would have been easy for Natives to transport and grow alders from seeds. Each tree can produce thousands of seeds each year, and they germinate abundantly. As my fieldwork has shown, alder seedlings of this species are very rare, because in order to germinate, they need wet, sunny gravel,[19] which has been rare since the glaciers retreated. Any place that is wet enough for alder seeds to germinate will also be shaded by larger trees. But Natives might have easily figured out how to germinate the seeds. They certainly would have had the motivation, since alder bark decoction is pretty good at controlling skin infections.

Holly for a Ceremonial Drink

The yaupon holly (*Ilex vomitoria*) is a shrub with ceremonial importance to several tribes, including the Cherokee. As the scientific name implies, it induces vomiting and clears out a warrior's digestive system prior to going to battle. The high levels of caffeine in the leaves might be the reason. Leading women, like Nancy Ward (Elizabeth's great-grandmother), had the responsibility to mix up the "black drink," with yaupon leaves and salt,

138   *Forgotten Landscapes*

for the warriors. The important Cherokee towns, in the Appalachians, are outside the natural range of the yaupon, which grows primarily on the Gulf Coast. Presumably the Cherokees planted the yaupons near their towns.

Bois d'Arc for Wood

The bois d'arc tree (*Maclura pomifera*) is a relative of the mulberry. Like other mulberries, it has separate male and female trees. The female trees produce fruits that look like big, sticky green mulberries because that is what they are. While not poisonous, bois d'arc fruits are not something you would want to eat, particularly because the latex might glue your teeth together. I, for one, am not willing to eat one of them to find out. As a matter of fact, there are no native animal species that regularly eat bois d'arc fruits and swallow the seeds. Until the end of the last ice age, there were numerous large mammal species, such as mammoths, mastodons, and gomphotheres, as well as native horses, that might have eaten the fruits. But once those species became extinct, there were no animals that would eat the entire fruit. I have often seen squirrels tear open the fruits and eat the seeds, but this kills the seeds rather than dispersing them.

The extinction of the large mammal seed dispersers makes the bois d'arc an example of what some scientists consider to be evolutionary anachronisms.[20] Bois d'arc was dependent on the large mammals for its effective seed dispersal, but now only the bois d'arc remains, like a waltz with no partner. I have seen bois d'arc seeds germinate even without scarification, but when they do so, all the seeds from a fruit germinate in a dense clump, in which not only most of the seedlings die, but they never fall very far from the parent tree—that is, unless they fall in a river. Whether or not this means that, eventually, this species would have become extinct when the last fruit rolled from the Mississippi into the gulf we can never know, because it has been rescued from extinction: It was extensively planted as windbreaks and livestock barriers throughout North America. Before the invention of barbed wire, the thorns assured that only the most determined cow could escape through a bois d'arc hedge.

The evolutionary anachronism theory overlooks the possible role of Native Americans in transporting the bois d'arc seeds to new locations. They highly valued the extremely strong wood of bois d'arc, especially for making bows. (*Bois d'arc* is French for "wood of the bow.") They could have planted the seeds in new locations to have a ready source of wood for making bows. This, however, was unlikely; archaeological evidence suggests that it was the wood itself, rather than the seeds, that traveled on trade networks. The wood probably (but not necessarily) came from wild trees.[21]

Coffeetrees for Game Pieces

The Kentucky coffeetree (*Gymnocladus dioicus*) is a large leguminous tree that grows widely scattered in the Eastern deciduous forest. It is so called because desperate Europeans and white Americans roasted the fruits of the female trees to produce a coffee substitute.

The distribution pattern of the coffeetree is puzzling. While it grows widely in Eastern forests, it is rare wherever it grows. Though the unroasted seeds are inedible, Bonnicksen has suggested that Natives carried the large, dark, shiny seeds with them as game pieces.[22] They would certainly have been lighter and more uniform in size and appearance than pebbles, making them more effective as game pieces. Once in a while, the players would lose some of the seeds, which could then germinate and grow into a tree. Once a tree got established, it could fill an entire area with underground stems.

There is no way to prove this explanation for the incredible journey of the coffeetree. Is it too much to suggest that each widely spread clump of coffeetrees may have begun when a pre-Columbian Native lost some of his game pieces?

Sugar Maple for Sugar

My next example is the sugar maple (*Acer saccharum*). One can hardly think of New England without thinking of maple syrup. One wonders how it was ever discovered, because the sap that rises in the spring must be boiled down forty to one to make the syrup. It was the Native Americans who figured out how to make maple sugar, which was almost the only sugar they had, and was undoubtedly an important trade item. The white New Englanders learned it from the Natives. Sugar maple is abundant. But is it too much to suggest that the range and abundance of sugar maples was enhanced by Natives carrying the seeds, which can be easily collected from the ground, to new locations?

Honey Locusts for Food and Game Sticks

My final example is the honey locust tree (*Gleditsia triacanthos*). Cherokee historian James Mooney, in 1900, indicated that the Cherokees sometimes planted honey locust orchards. They used the sweet pods as a source of food and medicine, and the wood for game sticks. Honey locusts today have a very widespread distribution, even though the fruits and seeds travel only very slowly. Within any location, underground stems seem to be the predominant manner of spreading.

140 *Forgotten Landscapes*

Some have identified the honey locust as one of the trees with "nonsensical" fruits most likely spread by extinct large herbivores.[23] But another dispersal agent has been suggested by a different author: humans, in particular Cherokee Natives.[24] The evidence one author considers most compelling is that some honey locusts grow in the wild in locations where the physical conditions are not optimal for seed germination and seedling growth (they are too wet), but near former Cherokee habitations. However, honey locusts spread primarily by underground stems rather than seeds.

Other Examples?

Native Americans utilized literally hundreds of different plants for medicinal purposes. One example is the slippery elm (*Ulmus rubra*), which produces a mucilage in its inner bark that is valuable, among other things, for dressing wounds. Natives used it for this, and whites adopted the procedure. One of Andrew Jackson's duel wounds was dressed with it. However, in most cases, Natives could easily have gathered the material they needed from wild populations, especially an abundant species such as slippery elm. On the other hand, it would be equally difficult to prove that they did not carry any seeds of medicinal plants with them to plant in new locations. I merely mention here that there may be other examples that have simply not occurred to us.

Table 6.1 summarizes examples of plants whose distribution may have been deliberately enhanced by Native Americans.

NEW HOPE FOR MODERN ORCHARDS

Katharine Lee Bates, inspired by the scenery from the top of Pikes Peak in Colorado, wrote the words for "America the Beautiful." At least in earlier decades, nearly every American school child sang this song, with her famous words about spacious skies, amber waves of grain, purple mountain majesties, and the fruited plain. She attributed credit for this beauty to America and to its God ("God shed His grace on thee").

While this may be true of the skies and mountains, the amber grain was not natural. It was the product of industrial agriculture. In eastern Colorado at that time, she wrote, there was enough rain to raise wheat without irrigation, and without the need for careful conservation of the soil. When

Table 6.1.

Edible Fruits	
Blackberry	*Rubus* sp.
Black cherry	*Prunus serotina*
Chokecherry	*Prunus virginiana*
Crabapple	*Malus* sp.
Currant	*Ribes* sp.
Elderberry	*Sambucus canadensis*
Fan palm	*Washingtonia filifera*
Grape	*Vitis* sp.
Mulberry	*Morus rubra*
Passionflower	*Passiflora* sp.
Pawpaw	*Asimina triloba*
Persimmon	*Diospyros virginiana*
Plum	*Prunus* sp.

Edible Nuts	
Beech	*Fagus grandifolia*
Chestnut	*Castanea dentata*
Hickory	*Carya* sp.
Oak	*Quercus* sp. (primarily California)
Pecan	*Carya illinoensis*
Piñon pine	*Pinus edulis* and *P. monophyla*
Walnut	*Juglans* sp.

Other Uses	
Bois d'arc	*Maclura pomifera* (wood)
Coffeetree	*Gymnocladus dioicus* (hard seeds)
Honey locust	*Gleditsia triacanthos* (sugar and wood)
Maple	*Acer saccharum* (sugar)
Seaside alder	*Alnus maritima* (antibiotic)
Yaupon holly	*Ilex vomitoria* (ceremonial)

droughts began in the late 1920s, the grain died and the soil blew away, creating ecological devastation. The "fruited plain" which she saw was the peculiarly white American orchards, which were also devastated by the Dust Bowl. Today, agriculture survives in eastern Colorado because of massive pumping of groundwater from the Ogallala Aquifer, which has been sucked so dry by pumps that each drop of water is expensive. Perhaps if the fruited plain had been Native fruits that could and did survive the dry conditions—such as sand plum and sand cherry, fruits still popular for making jellies and jams—the orchards of the fruited plain might not have died. And the farmers might have had Native Americans to thank for it.

7

EUROPEAN DISEASES

The Beginning of the End of Native America

Given all of the ecological and economic benefits of Native food production and landscape management, and their strong healthy populations, there would seem to be no reason that Native cultures could not have continued to flourish after European contact. The Vikings could not conquer them. Even after the collapse of Mississippian civilization, they presented a strong resistance to Europeans upon initial contact. Strong Native American cultures might have continued indefinitely into the future. They did not, however, for two reasons.

First, Europeans and white Americans began a long and sordid process of slaughter, slavery, and rape against Native Americans, of which I have no intention of giving details in this book. This is, however, the same thing that Europeans did in many other places in the world, from Africa to India to China. Despite this, African and Asian cultures continue to play a major role in the world, unlike Native American cultures.

Second, the Europeans brought diseases to which the Natives had no resistance. These diseases spread rapidly in Native communities and had devastating effects on Native populations. Since nobody is certain of how many Natives were present before the European plagues, nobody knows how many of them died. Estimates range from 50 to 90 percent. This is why modern America is mostly white. Seldom, perhaps ever, has the losing side of a conflict suffered such losses on a continental scale. This did not happen in Africa, India, or China. It took centuries for the European colonists to conquer the Native Americans; if there had been ten times as many of them, the Europeans might not have succeeded. America, under a different name, would still have a place on the world stage, but it would be a lot less white. America might be like modern Africa, India, and China, where whites are a minority.[1]

144 *Forgotten Landscapes*

Massive deaths of Natives due to disease is documented both by European and Native observers. Examples of diseases that killed overwhelmingly more Native Americans than Europeans include such serious diseases as smallpox, bubonic plague, malaria, scarlet fever, gonorrhea, typhoid fever, typhus, and an Old World form of tuberculosis.

The decimation of Native Americans by Old World diseases was far from the only example of disease changing the course of human history. Some historians attribute many of the major turning points of world history to epidemics. The Plague of Athens, in 430 BCE, devastated the Athenian population and, according to some, contributed to its eventual defeat by Sparta, which was a less philosophical and more military city-state.[2] The Fall of Rome may have been strongly influenced by one of the many plagues that occurred at that time.[3]

Many important questions in world history cannot be adequately answered without taking diseases into account. For example, why is Britain mostly English? The Roman Empire had taken the British Isles from Gaelic tribes. But when the Romans left, the Gaelic tribes reached an unstable peace with Angles and Saxons that had migrated from what is now Germany. When a plague hit the Gaelic tribes after 541 CE, the Angles and Saxons were able to overcome them. Today they are the English.[4]

The most famous example of a European plague is the Black Death of 1347. Bubonic plague bacteria entered Mediterranean Europe in 1347, and from there it spread throughout Europe, reaching Scandinavia in 1350. It killed one-third of Europeans. So many people died that the entire feudal economy nearly collapsed. Serfs had previously been tied to the land, but after a third of the serfs died, the survivors said, in effect, "Yeah, we'll plow and harvest, if you pay us." People had their faith in the Catholic Church shaken, since the Church (in which a large number of priests perished) was unable to predict or prevent the Black Death. With younger priests filling positions vacated by the deaths of older priests, there was a little more room for new ideas. Reform movements, eventually the Reformation itself, may have been the result.

New waves of plague continued to afflict Europe for hundreds of years. Plagues of diseases such as typhus killed more soldiers and civilians than did actual combat or slaughter during the interminable civil wars of Europe. Many books, one of the earliest of which was Hans Zinsser's *Rats, Lice, and History*, have been written about the historical effects of European diseases.[5]

The massive mortality of Native Americans due to European diseases, however, was astonishing (50 to 90 percent), even by Black Death

European Diseases 145

standards (about 30 percent). It is instructive to compare the American experience with a wave of epidemics that occurred only a little later: plagues of European diseases killing Africans. But the outcome was strikingly different. Africa is still overwhelmingly Black. Uncountable masses of people, white and Black, died of disease, but they were often the same disease, such as smallpox. Europeans conquered the Africans but did not replace them to the extent that they replaced the Native Americans. For every European disease killing an African, there was an African disease to kill a European. Being sent in a European army to Africa was considered virtually a death sentence.[6]

The result of the massive Native American depopulation is that European immigrants found not only a rich land—transformed by Native fires and agriculture into a landscape ready for habitation—but one that was virtually empty, one which its surviving Native inhabitants could not defend. As a result, the European immigrant populations exploded—in fact, serving as a textbook example, according to Thomas Malthus, considered the father of population studies.

Malthus literally wrote the book about population explosions, from which both Charles Darwin and Alfred Russel Wallace got their ideas for natural selection. Malthus got his numbers from English colonial census figures. He produced actuarial tables and calculated that, for every 100 deaths, American colonists had 300 births. In contrast, back in Europe, where almost all the suitable land was already in production, there were only 117 births for every 100 deaths. To Malthus, and to modern economists, the European situation is what usually happens, while the American situation happens only under special conditions of sudden and almost limitless resource availability. Adam Smith, often considered the father of modern economics, also compared America to Europe. He said that while the white population doubling time was five hundred years in England, it was only twenty-five years in North America.[7] The immigrants overwhelmed the depopulated Natives and turned them into little tribes, many of which are barely clinging to cultural survival today.

No one, either invader or Native, understood why these plagues occurred. Occasionally, the white conquerors attributed them to God, who was allowing them to conquer the heathen Natives. In this famous quote, King James I said of the plagues in seventeenth-century Massachusetts,[8]

> Within these late years, there hath, by God's visitation, reigned a wonderful plague, the utter destruction, devastation, and depopulation of that whole territory, so as there is not left any that do claim or challenge

146   *Forgotten Landscapes*

any kind of interest therein. We, in our judgment, are persuaded and satisfied, that the appointed time has come in which Almighty God, in his great goodness and bounty towards us, and our people, hath thought fit and determined, that those large and goodly territories, deserted as it were by their natural inhabitants, should be possessed and enjoyed by such of our subjects.

The colonists sometimes made similar statements. A former South Carolina governor said in 1707 that it was God's will "to send unusual Sicknesses [to the tribes] to lessen their numbers; so that the English, in comparison to the Spaniard, have but little Indian blood to answer for."[9] But many colonists were careful to not praise the epidemics, which killed the colonists in large numbers also.

SMALLPOX

Perhaps the most dramatic of these diseases was smallpox. No one is certain where it began, but it was an Old World disease. It produced symptoms that could hardly be mistaken for any other disease. Among the many symptoms were the skin pustules themselves, which are numerous, filled with pus, surrounded by an inflamed area of skin. They could be so large and numerous as to become confluent—that is, run together into giant pustular lagoons. The victim also had extreme respiratory symptoms, in part because the pustules also occurred on internal membranes. The viruses produced toxins that made the victim feel very sick. A worse way to die could hardly be imagined. Europeans recognized it instantly, and Natives knew they had never seen any such disease. The aftereffects of smallpox, in those victims who happened to survive, were also devastating, their skin left disfigured forever by scars.

Table 7.1 lists just a few examples of smallpox plagues that killed large percentages of Native tribal populations. These percentages cannot be calculated, and I have not listed them. One can see from these few examples that deaths by disease were rampant all over Mexico and the North American continent, and continued for over three hundred years.

The worst was in Mexico, where within a half-century the Native population fell from twenty-five to thirty million to a mere three million. Some estimates put the total number of Native American deaths at fifty-five million.

A similar thing was happening in the Amazon rainforest during the same time frame. This rainforest is inhabited by "primitive" tribes, some

Table 7.1. Examples of major smallpox plagues in Native populations.

Year	Tribe or Location
1519	Mexico
1546	Mexico
1613	Quebec
1633	Massachusetts
1639	Quebec
1738	Cherokees
1762	Ohio
1780	Plains tribes
1837	Mandans
1862	Pacific Coast

of which remain uncontacted by the outside world directly, although they know about the outside world because of their contacts with other tribes. Many people believe that the "naked savages" living in temporary huts represent the timeless condition of these tribes. That's certainly the image you get from movies such as *Emerald Forest*. But recent evidence, some on the ground[10] and some from airborne laser imagery sponsored by the French Centre National de la Recherche Scientifique,[11] has demonstrated that the Amazon Basin had some large cities and roads, which disappeared before the first whites reached the interior of the forest.

Not only did these "primitive tribes" build structures, but they also changed the landscape. Some tribes intentionally used composting to improve the quality of the notoriously poor rainforest soil. It may have been European diseases that chased previously civilized tribes from their cities into the rainforest.[12]

Why You Don't See Smallpox Anymore

If you've never seen anyone with smallpox scars, you have the World Health Organization (an agency of the United Nations) to thank for it. (In January, 2025, the United States began the process of withdrawing from the World Health Organization.) The successful eradication effort is described in detail in the "Red Book" edited by Frank Fenner.[13] After the United Nations formed at the end of World War II, it looked for an important project, but one limited enough that they could accomplish it. They chose to eradicate smallpox. All of it, everywhere. They began programs of mass immunization. Their health workers went into every local village that still had smallpox, mostly in Asia and Africa. They didn't have to immunize

148   *Forgotten Landscapes*

everyone, but just enough that smallpox was unable to spread. Since the smallpox virus lives only in humans, the health workers could eradicate smallpox by drastically reducing the number of *susceptible* human hosts.

The last person to get smallpox by natural transmission was a Somali cook in 1978, who later recovered. Another person died in England during a lab accident in which viruses got into the ventilation system. By 1980, the World Health Organization declared smallpox extinct. A few vials of virus remain in research laboratories, where, it is assumed, they will remain, although the Soviet Union did research on using smallpox as a biological weapon in the 1980s.[14]

Smallpox Comes to America

Upon European contact in the 1500s, smallpox spread rapidly throughout the American continent. One reason is that the viruses are very infectious and constantly emerge from pustules and from particles coughed out by the victim. But the main reason smallpox spread rapidly is, in fact, the network of trade routes I described in chapter 1. A Native infected with smallpox would be able to travel quite a distance before feeling sick. It did not take very long at all for smallpox to spread throughout the continent.

The disease certainly spread faster than did the Europeans themselves. Europeans mostly saw inland villages only after smallpox had devastated them. Smallpox reached the Native villages of the Pacific Northwest before the first white contact. When Captain George Vancouver met Natives near Puget Sound in 1792, he found empty villages and a few pockmarked survivors. The disease must have come over Native trade networks, since only a few Spanish captains had preceded him, and then only by a year or so. The greatest Native societal strength—their trade networks—also proved to be their greatest weakness to diseases. A continent-wide smallpox epidemic about the time of the American Revolutionary War spread by trade among both Europeans and Natives. Trade networks bring death as well as life.[15]

Smallpox was just one of many diseases that afflicted the Europeans and which the Native Americans had never encountered. The Native Americans, in contrast, had very few diseases of their own. There are several possible reasons for this. As mentioned in chapter 1, Native cities and villages were clean. The Natives were also well-fed and healthy, in contrast to many poor Europeans. Infectious diseases are less likely to originate among well-fed and physically strong people. Native societies were not the kind that would either generate or maintain their own epidemic diseases. It is also likely that the ancestors of the Natives brought few diseases with them across the Bering Land Bridge.

European Diseases   149

Another important reason Natives had few epidemic diseases is that Natives did not have livestock. Several major Old World human diseases emerged from livestock species.[16] Smallpox is itself an example. Smallpox (variola) is very similar to cowpox (vaccinia), so similar in fact that the human immune system reacts to cowpox as if it is smallpox. Immunity against the mild cowpox makes a person immune to smallpox. This is the basis of the first vaccination carried out by Edward Jenner in England in the late eighteenth century, who immunized people against smallpox by exposing them to cowpox. In modern times, cowpox has been a mild disease for both cows and humans. But smallpox might have begun as a deadly variant of cowpox, and spread from cows to humans. Tuberculosis apparently also began as a livestock disease.

An Example: Smallpox in the Cherokee Tribe

The smallpox plague that struck the Cherokee tribe in 1738 was not the first or the last such plague in the tribe, but it was the most dramatic. This plague killed about half of the people in the tribe, irrespective of age, gender, or state of health. The Cherokees knew the Europeans had brought the disease, but did not know how to stop it. Families and communities were disrupted by the unpredictable massive deaths. The Cherokees were unable to maintain their previous levels of agriculture and hunting, and the strong trade network was now a source of fear rather than of strength. Cherokee religious leaders attributed the plague to moral degeneration, including sexual interactions with, and rum from, the white traders.

According to James Adair's 1775 account, even the survivors were thrown into depression when they saw themselves disfigured by the disease:[17]

> A great many killed themselves; . . . some shot themselves, others cut their throats, some stabbed themselves with knives, and others with sharp-pointed canes; many threw themselves with sullen madness into the fire, and there slowly expired, as if they had been utterly divested of the native power of feeling pain.

Adair described in particular detail the death of one Cherokee smallpox victim:

> When he saw himself disfigured by the small pox, he chose to die, that he might end as he imagined his shame. When his relations knew his desperate design, they narrowly watched him, and took away every sharp instrument from him. When he found he was balked of his intention, he fretted and said the worst things their language could

150   *Forgotten Landscapes*

express, and shewed all the symptoms of a desperate person enraged at this disappointment, and forced to live and see his ignominy; he then darted himself against the wall, with all his remaining vigor,—his strength being expended by the force of his friends' opposition, he fell sullenly on the bed, as if by those violent struggles he was overcome, and wanted to repose himself. His relations through tenderness left him to his rest—but as soon as they went away he raised himself, and after a tedious search, finding nothing but a thick and round hoe-helve, he took the fatal instrument, and having fixed one end of it in the ground, he repeatedly threw himself upon it, till he forced it down his throat, when he immediately expired.

One of the survivors of the 1738 Cherokee epidemic was a young boy later called Tsiyu Gansini, or Dragging Canoe, whom I introduced in chapter 1. He resented the whites not only for taking Cherokee land and killing Cherokee warriors directly, but also for bringing disease. He wore red and black war paint on his face for the rest of his life, partly to cover his smallpox scars. He was the implacable opponent of all peace efforts between Cherokees and whites, especially those of his cousin Nancy Ward. By the 1780s, most of the Cherokee tribe had come to accept white domination. But not Dragging Canoe and his followers. They formed their own Chickamauga community up in the mountains of northern Georgia. He remained a formidable force, until the Chickamauga were devastated by yet another smallpox plague, this one in 1783. Dragging Canoe died in 1792, and his followers surrendered two years later.

Another Example: Upper Plains Tribes

Other tribes were nearly eradicated. A case in point is the Mandan tribe of the upper Midwest. In 1837, white people on a steamboat introduced smallpox to the tribes. No accurate estimate of the number of deaths was possible even at the time, but it is believed to have exceeded 17,000. Fur trader Francis Chardon wrote that only 27 Mandans were left alive.

WHY NATIVE AMERICANS WERE SUSCEPTIBLE TO OLD WORLD DISEASES

The question then remains as to why Natives were so helplessly susceptible to European diseases. The explanation that is usually provided is that Native American populations had no opportunity (if plagues can be called opportunities) to evolve resistance to them. For thousands, perhaps millions, of years, Old World human populations had encountered these diseases. The people who could resist them lived and reproduced; the ones who could not died, usually before having a chance to have children. Natural selection produced resistant Old World populations, not just in Europe but also in Asia and Africa.

In North America, this evolutionary process had not occurred. This was the basis of the idea of "virgin soil epidemics"—that is, diseases that cause epidemics in populations that had never encountered them. This clearly happened in America, and continued well into the twentieth century, in which previously uncontacted tribes in the Amazon rainforest died from what Europeans and white Americans considered minor diseases, such as influenza, chickenpox, the common cold, measles, and pertussis. This was a major premise of the novel *At Play in the Fields of the Lord* by Peter Matthiessen. This is what I taught my students over the course of decades: Native Americans died from diseases against which they had evolved no resistance.

I first learned about virgin soil epidemics in North America by reading Alfred Crosby's book *Ecological Imperialism*. He maintained that one reason Europe conquered much of the world was that European diseases killed so many victims.[18]

Balanced Pathogenicity

The process by which this occurred has been called *balanced pathogenicity*. Disease is an ecological balance between parasites (such as smallpox viruses) and hosts (such as humans). Upon the first encounter of parasite and host populations, most of the hosts have no resistance, and natural selection favors the few who, for whatever reason, do not die of the disease. This process of natural selection changes the host population. The resistant hosts become more common in the population, making the population as a whole more resistant. Meanwhile, the most successful viruses are the ones that can infect the most hosts.

But a successful virus is not always one that kills its host quickly and dramatically. Such viruses are called *acute*: They have severe effects upon the host and kill the host quickly. This may be a winning strategy for the virus upon first contact, but over the long term it is a bad one. The host is their world. By killing a host, the viruses are destroying their world. An acute virus can be severe but it is not the most successful. As a result, over time, natural selection favors the viruses that do not destroy their hosts

152   *Forgotten Landscapes*

quickly, or at all. The viruses become more *chronic*—that is, they have mild effects upon the host and may not kill it at all.

A chronic parasite can be more successful than an acute one. Not only does the acute virus destroy its world, but it scares away other possible hosts. This is what the Ebola virus does. The host dies quickly, bleeding from every orifice, and everybody knows it and stays away. A chronic parasite, in contrast, may make the host feel a little bit sick, but the host still feels good enough to get up out of bed and go to the market and, unknowingly, spread parasites to other potential hosts. The advantages of being mild balance the pathogenicity of the parasite.

I first learned about balanced pathogenicity from reading some of the works of microbiologist René Dubos.[19] I was, at the time, teaching a college microbiology course for which I was totally unprepared. I started off assuming that "germs" were always bad. Dubos's ideas came to me like a revelation. For him, it certainly was. He went on to write many books that suggested that evolution produced a world in which former enemies could become cooperative partners.

In fact, Dubos said, some diseases evolved so far in the direction of balanced pathogenicity that they were no longer recognizable as diseases. He gave examples of diseases that had, apparently, disappeared from human history. The "sweating sickness" afflicted many people in sixteenth-century Europe. As the name suggests, one of its main symptoms was profuse sweating. Within a couple of hundred years, there were no more cases. A disease with such obvious symptoms could hardly be overlooked even in the days before modern medical science. Dubos said that the parasite evolved into such a mild form that it seemed no different from colds and flus.

A very recent example of balanced pathogenicity and its alternatives is the worldwide COVID-19 pandemic. When the disease first appeared, it was dangerous and killed a lot of people, at least three million in 2020, according to the World Health Organization. People, especially older ones, would become unable to breathe without a respirator. The disease, caused by a coronavirus, was also highly transmissible. An immediate worldwide campaign to develop and administer vaccines was begun. To further reduce transmission, people stayed home unless they had to leave. They worked from home, and attended school from home, when possible. I became a talking head on a computer screen instead of in a classroom.

Over the next couple of years, new variants of COVID kept emerging. Each new variant would spread, while some older variants would become rare and perhaps disappear altogether. Because the vaccines targeted a part of the virus that does not change very much, previous vaccinations

European Diseases   153

helped to protect people from the new variants. But the newer variants had two characteristics that distinguished them from the original strains: First, they were milder. It got to the point where having one of the new strains of COVID was hardly recognizable as being something other than just a light influenza. Second, they were more transmissible. Although the new COVID variants were milder, it was almost impossible to not be exposed to them. COVID has evolved from being an acute to being a chronic disease. According to the BBC, "Hospitalisations and deaths from Covid-19 are markedly lower [in January 2024] compared to January 2023. Primary care physicians say they are finding it virtually impossible to distinguish Covid-19 symptoms from influenza without the help of a [genetic] test."[20]

Another example of balanced pathogenicity is syphilis. Today, it is a chronic disease spread by sexual contact. It can severely damage the host, but takes a long time to do so. And it does not cause epidemics. It apparently originated as a strain of yaws, caused by a related species of bacterium, and which had generally mild effects when it circulated in Native populations in the Caribbean area. It had evolved into a mild disease.

But when syphilis reached Europe, it became a raging and deadly epidemic disease among people—this time, European—who had developed no resistance to it. The conclusion seems inescapable: Columbus's crew brought it back with them. The evidence, while circumstantial, is hard to deny. The first European syphilis epidemic occurred in 1495, right after the return of Columbus's first crew, during a French attack on Naples. Many of the French troops were Spanish mercenaries, which could have included men who had sailed with Columbus. The brutal sexual exploits of Columbus's crew, though not of Columbus himself, are well documented. Columbus gave his approval for his soldiers to rape Native women. If this is indeed where the severe form of syphilis came from, it is the one shining exception to the pattern of European diseases ravaging Native American populations.

I had my own experience with balanced pathogenicity. Before an international trip when I was in high school, I had to be tested for antibodies (which I will explain later in this chapter) to various diseases, one of which was tularemia. This was a disease that, as I explain in chapter 8, is named after the county I grew up in. When the doctor told me I tested positive for tularemia antibodies, I said I didn't remember ever having it. He said it had become such a mild disease that I might have had it and just felt a little off for a few days.

Smallpox itself is a product of balanced pathogenicity. In parts of the Old World, balanced pathogenicity had produced some hosts that were

154   *Forgotten Landscapes*

resistant, and some viruses that were mild. They were still variola smallpox, but now went by the name of variola minor. Variola minor, just like the dangerous variola major, spread to the Americas. My mother said that my grandmother (in our Oklahoma Cherokee lineage) had suffered from smallpox. She was very ill for several weeks, but then recovered, without scars. It was variola minor, produced by balanced pathogenicity, that had struck her.

When Balanced Pathogenicity Doesn't Happen

But in some cases, balanced pathogenicity does not occur during the evolution of diseases. If a virus is too acute, it will destroy its host. But if it can spread to new hosts quickly enough, and become more contagious, the death of the old host does not matter to the virus's success. This is what Ebola does. It can infect a new host even before the old host shows symptoms. Enhanced contagion is also the new face of COVID.

Balanced pathogenicity occurs most importantly when the parasite spreads to another host by direct contact or through respiratory droplets. The uninfected hosts can learn to recognize and stay away from sick ones. But what if the germ spreads without direct human contact? Some of the most important examples are waterborne diseases and diseases that spread by insects.

An example of the former is cholera, a disease that is very rare today. You usually get it by drinking or washing food in infected water. In the days before water treatment, you could get cholera because you had no idea where the water came from, such as an upstream village. The chain of transmission can be broken by interrupting the supply of water from an infected source, such as the Broad Street pump in London in the nineteenth century. To control the cholera outbreak, physician John Snow said to take the handle off of the pump.[21]

Spreading by water, cholera has remained a very virulent disease. But water supplies are not the only means by which cholera can spread. It can also spread by direct human contact. When it spreads by direct contact, as is the case with most other diseases that spread in this fashion, cholera can evolve to become milder. This can only happen if it is forced to happen. Microbiologist Paul Ewald and his team interrupted the transmission of cholera during a South American epidemic by supplying people with clean drinking water. As a result, the cholera bacteria had to evolve to be milder if they were to spread at all. He demonstrated that, in fact, the new cholera bacteria were milder than the original ones. He referred to this as *domesticating* the bacterium. This may be the most famous case of balanced

European Diseases 155

pathogenicity being put to work in the service of human health. But note that it only worked because public health officials forced it to.[22]

Finally, we consider the diseases that are spread by insects, such as mosquitoes, ticks, and lice. The world's most important example is malaria, which kills millions of victims each year. These diseases frequently evolve to become more acute, not less. The reason is that if the host is in bed, very sick, he or she is much less likely to run away from or swat the insects. A very sick victim is more, not less, likely to be visited by another insect.

HOW THE IMMUNE SYSTEM WORKS

At the same time I taught about balanced pathogenicity in my general biology classes, I also taught about how the human immune system worked. It was a pretty routine topic, at least until a significant minority of people began to dislike the idea that the government could encourage and subsidize immunizations. On the political left, there have always been back-to-nature people who have rejected immunizations. On the political right, conservative religious groups have considered immunizations to be against the will of God. It got worse during the COVID-19 pandemic. One of the few times then-president Donald Trump got booed was when he took credit for the rollout of "Operation Warp Speed," providing immunizations against COVID-19 to everyone. One of my students, from a rural conservative family, started off skeptical about COVID immunization, but then took it very seriously when his uncle and great-uncle, both refusing vaccinations, died within a week of one another.

Had I thought about it, I would have realized that what I taught about virgin soil epidemics partly contradicted what I taught about the immune system. The immune system is based on antibody molecules, which are part of white blood cells. Each white blood cell has just one kind of antibody molecule. Antibodies stick to antigens, which are molecules in the environment. Most antigen molecules are of biological origin. A lot of them are harmless, but many of them are associated with parasitic disease organisms. Antibodies are very specific. Each kind of antibody sticks to just one kind of antigen, and that's it.

Each white blood cell, therefore, is an extreme specialist. If it encounters its antigen, it sticks to it; if not, it slips on past. The only way a white blood cell can recognize its antigen is by sticking to it. This is how antibodies—and therefore, white blood cells—can get rid of disease antigens. So many antibodies stick together with antigens that they form

156   *Forgotten Landscapes*

clusters which other cells can consume and destroy. This is obviously a very oversimplified explanation.

When a baby is born, he or she may have white blood cells that contain up to one hundred billion different kinds of antibodies, each white blood cell having just one kind.[23] This is surprising, since the genetic basis for making these antibodies consists of only a small number of genes. But while the white blood cells are developing, the antibody components get cut and spliced in myriad different ways. Thus, at the beginning of his or her life, a baby has about one hundred billion different kinds of white blood cells skulking around in his or her blood vessels. These white blood cells operate individually or in small groups, sort of like guards on patrol. The guards slip right past anything that does not have its antigen. There are few of each kind of white blood cell.

But if a white blood cell encounters its antigen, possibly a germ, everything changes. The white blood cell not only adheres to the germ but also starts to reproduce. There might have been a hundred of them to start with, but quickly the number swells to millions. Each of these cells is now part of an army that focuses on just one antigen—that is, on just one kind of germ. Meanwhile, the germ is also reproducing. The race is on. You'd better hope that the white blood cells reproduce faster than the germs.

After the antibodies win the battle against a disease, the body cannot afford to keep a full standing army of them. It begins to break some of them down. What starts off as individual patrols becomes a large army, then becomes smaller platoons. If the white blood cells encounter the same antigen again—which is likely, in any given environment—the immune response does not need to begin from a small number of white blood cells, but from an intermediate number. They respond to the germ before the person even knows he or she has been exposed. This is immunity. Vaccination is simply administering the antigens without the germs—that is, with the germs inactivated, to create an artificial immunity. A white blood cell can't generally tell the difference between a live germ and a dead one. If you get smallpox and recover—or get a vaccine—you are immune for years, or perhaps for a lifetime.

The diversity of white blood cells results from cutting and pasting the molecules into different combinations. Most people everywhere in the world start off with more or less the same building blocks of antibody parts. This means that anyone in the world, regardless of evolutionary history, is *potentially* capable of responding to any antigen. Therefore, there is no reason to believe that Native Americans had an intrinsically or genetically inferior ability to respond to smallpox antigens.

In other words, the immune system seems almost infinitely flexible. This is the only way that animals can survive in an environment that is literally swimming in parasitic threats. Almost any of them could kill you were it not for your skin and immune system protecting you. And not only that, but new parasitic diseases are evolving all the time. A disease agent needs to have only slightly different antigens to evade your body's existing antibodies.

Why Europeans Were More Resistant to Smallpox than Native Americans

When a European baby was born, it was helpless against smallpox until the immune reaction kicked in. Frequently the reaction was inadequate, and the child died. The survivors in the population were children and adults whose immune reaction against smallpox worked. But smallpox remained an extremely frightening and dangerous disease for Europeans. They were not born immune to smallpox. They became immune, if they were lucky. Lots of Europeans were not lucky. The scarred skin left by smallpox infection was a common sight in Europe.

One European who was not lucky was Lady Mary Wortley Montagu. She was very beautiful. Then she got smallpox. She recovered but was ashamed of her scarred appearance. She wanted to find some way that she could help save the lives of her fellow Europeans. Her husband was the British ambassador to Turkey. Lady Montagu spent a fair amount of time in Turkish bathhouses where she saw the Turkish women without clothes. She was astonished that none of them were disfigured by scars. When she inquired about it, she discovered that most Turkish people were artificially inoculated with pus from a smallpox victim. Nobody knew about germs, but there was something about the pus that conferred immunity when administered to human skin.

When she returned to England, Lady Montagu began a campaign to have as many people as possible inoculated with smallpox pus.[24] To say she met with opposition is an understatement. The process did not always work. An inoculated person could catch the disease itself rather than become resistant to it. But eventually she gathered enough allies and they convinced enough people that inoculation became common in England. It was slower to catch on in France. Inoculation (also called variolation) was known throughout Asia and Africa. Everybody in the Old World except the Europeans seemed to know about it. Doctors got better and better at administering the inoculation, eventually reducing the risk of death to about 1 percent.

158   Forgotten Landscapes

The Europeans who came to America were often those who had smallpox when they were younger, and were immune by the time they reached the shores of the New World. Many others had been variolated. Smallpox was still, for them, a very dangerous and frightening disease. Meanwhile, the Native Americans, even starting with the same immune building blocks, had never been exposed to the virus antigens. When they encountered the virus, their immune systems had very little time to respond. They were all as helpless as newborns. Today, Native Americans do not appear to be any less capable of developing immunity to any disease to which whites can become immune, once differences in general health and access to health care are taken into account.

Europeans were often immune to smallpox, Native Americans, almost never. Europeans whose ancestors had come to America had intermediate levels of immunity. This happened because they lived at lower population densities, at least out in the country, than Europeans. American-born Europeans were exposed to smallpox much less often than Europeans in Europe. As a result, European Americans had every reason to dread smallpox plagues.

When ships arrived in Boston harbor with smallpox victims aboard, during the seventeenth century, the ship was quarantined. (The term *quarantine* comes from the Latin-based word for "forty," reflecting the number of days that Jesus fasted in the desert.) Quarantine helped to reduce the arrival but not the local spread of smallpox in Boston. A physician, Zabdiel Boylston, tried to variolate as many people as possible, in return for which crowds gathered and vandalized his home. Political leader Cotton Mather met with similar resistance.

A tipping point came when General George Washington noticed that many of his troops died of smallpox, more so than the British troops, many of whom had been variolated before they shipped over to America. Eventually Washington started requiring variolation, and the results might have made the eventual difference in the American victory in the Revolutionary War, particularly as there was a lot of smallpox at the final battle of Yorktown.

Rich American plantation owners relied heavily on slave labor. But slaves, living under unhealthy conditions, were at risk of catching smallpox. Some slave owners, such as Thomas Jefferson, decided to undertake the expense of having his slaves variolated. Jefferson had vials of pus delivered to Monticello so that he could variolate his servants, often doing the process himself. When he took some of his slaves to France during his ambassadorship, he had them variolated also, including his favorite slave, Sally Hemings.[25]

European Diseases 159

Virgin soil epidemics are merely the dramatic tip of a big iceberg. Almost anyplace you go in the world, even if the standards of public health are similar to where you come from, you can expect to be sickened by relatively minor diseases to which the natives have already developed resistance.

HOW AMERICAN EPIDEMICS CHANGED
AMERICAN—AND WORLD—HISTORY

Every part of the world has experienced plagues that have killed thousands or millions of people. The New World was spared from this experience until 1492. Starting at that time, Old World diseases have caused epidemics that have killed up to 90 percent of many Native American populations. This is the reason that North America is mostly white, while Africa is still mostly Black and Asia is mostly Asian. Only in America did the Natives have almost no diseases (with the exception of syphilis) that could strike back at the white conquerors.

It is possible that eventually white firepower would have overcome Native American defenses. Jared Diamond speculated that it was not just germs, but also guns and steel, that delivered Native Americans into the hands of European conquerors.[26] But it would have taken longer, and the conquest would have been less complete. Maybe at least some American diplomats to Europe would have been Native rather than white. They might even have spoken Cherokee. Thinking back to the transatlantic crossings of Attakullakulla in 1730 and Ostenaco in 1762, as described in chapter 1, it would not be the first time this happened.

8

A TOXIC PARADISE

The San Joaquin Valley as the Ultimate White Monoculture Dream

When I grew up in the San Joaquin Valley, which is the southern half of the Central Valley, I thought it was the most beautiful place in the world. Huge tracts of land were covered by orange trees, whose evergreen leaves had a pungent citrus smell and whose flowers filled the air with heavenly sweetness every April (now, every March) (figure 8.1). The few inches of rainfall occurred only in the winter, but all summer long the canals of the Central Valley Project (chapter 5) glistened blue (figure 8.2).

Figure 8.1. Orange blossoms filled the San Joaquin Valley air with an incredible citrus perfume every spring, and the trees (with dense evergreen leaves) produced a bounty of large, sweet oranges every autumn. *Author photo.*

162   Forgotten Landscapes

Figure 8.2. The Friant-Kern Canal continues to bring fresh water to parts of the San Joaquin Valley through the long dry summer. The water is melted snow and has few minerals. *Author photo.*

And on a few days each winter, after rain in the Valley and snow in the mountains, I could see the blue, snowy peaks of the Sierra Nevada mountains beyond the smooth foothills to the east.

The San Joaquin Valley is the southern half of the Central Valley, a broad flat plain between the Sierra Nevada and the coastal mountains of California. You cannot see all the way across. Through much of the last million years, the valley was the bottom of a vast lake of glacial meltwater. Because of this, much of the Valley west of the Sierra foothills is totally flat (flatter than Kansas, and that's a fact). It also has many layers of very rich soil because of multiple deposits of soil from the mountains. Because of its rich soil and virtually frost-free winters, you can grow almost every kind of crop and orchard imaginable. As a newspaper editor told me, "If you can grow it, you can grow it here."

I grew up in the little town of Lindsay. A more pleasant location than Lindsay—and other nearby Tulare County towns—is difficult to imagine, if you wanted to live in a garden. If you believed that nature could not reach its full potential unless it was subjugated by the Hand of Man, the San Joaquin Valley was the place for you to be. I would often ride my bicycle up one of the rural roads to the flank of a nearby hill to the north. From there, I could see far into the distance (figure 8.3).

To the east of Lindsay, the Sierra Nevada mountains—with Mount Whitney at their crest—towered higher than even the Rockies. On clear winter days, while the other kids played tetherball, I would sit on the playground and make sketches of Mineral King (the site of a failed Disney ski resort) and other peaks of the western ridge. Mount Whitney itself was on the higher, eastern ridge on the other side of the Kern River basin, where we could not see it.

Many species of giant coniferous trees, from huge-coned sugar pines to red-barked incense cedar, filled the Sierra forests. Emerald-green meadows lined some of the creeks. The forests were not only beautiful but

Figure 8.3. Citrus trees fill the San Joaquin Valley floor, while the foothills have pasture. *Author photo.*

164   *Forgotten Landscapes*

contained groves of giant sequoias, the largest trees in the world (chapter 2). Hidden in secluded coves, the sequoia forests were remnants of ancient woodlands that once covered much of the Northern Hemisphere. These long-lived trees stored thousands of tons of wood in their trunks.

These forests were my image of Heaven. I said this when I preached a sermon at a Boy Scout Camporee in the Sierras when I was a kid. But even to people less inspired than I was by the trees, the Sierra forests rewarded us with fresher and cooler air than we had thought possible down in our smothered valley.

Meanwhile, back down to the hill north of Lindsay, where above the orchards I could see some of the most carefully tended agricultural plots in the world. California was a frequent destination for Japanese families after the end of the feudal era, around 1870. Japan was a world leader in producing lots of food from intense gardening on hilly slopes. The first-generation Japanese (Issei) brought their skills with them to grow vegetables when they moved to the San Joaquin Valley, while the second generation (Nisei) continued the tradition. They treated their tomato patches so carefully that, in the springtime, after the daytime air was warm but the nights could still be chilly, they covered each plant with a white paper "hotcap." The hotcaps formed perfect rows on the hillsides until the gardeners removed them in April.

To the south and west of the hill where I stood stretched many square miles of orchards and other farmlands. And I mean square. Each road (east–west) and avenue (north–south) enclosed square-mile blocks. Orange trees covered much of the land. There is no natural forest that looks like an orange grove. All the trees within each orchard were genetically uniform. All were the same size. They had even been trimmed to exactly the same height and width. I knew a man whose livelihood was to trim away any orange tree branches that rose above a certain height. As a result, all the oranges grew on the outside of the almost-hemispherical trees at the perfect height for migrant farmworkers to access them with ladders. The trees were all trimmed down to an even greater uniformity than you would see in a flock of sheep. There were also orchards of silvery-leaved olive trees. Less often, you would see orchards of almond and plum, which blazed with white and pink in the early springtime before the leaves emerged. These orchards hid most of the rural houses, except when an occasional palm tree or two-story house emerged from the ocean of branches.

The highest mountains received massive amounts of snowfall, the kind that buried the Donner Party in 1847. The snow fell during the winter and melted slowly over the summer, creating numerous rivers that did not dry up even during summer drought. By the time I lived there, the Bureau of Reclamation and the Army Corps of Engineers had already built huge dams to trap the snowmelt into large reservoirs. Even these reservoirs could not hold enough water to quench the summer thirst of the Valley. The most important way that water was stored for human use was the snowpack itself. We all counted on the snowpack to gradually melt and provide us with water all summer. And, when I lived there, it always did.

When the first Spanish (later Mexican) and American explorers visited the Valley, they saw vast expanses of dry prairie. Modern scientists understand that the grasslands were maintained, and shrubs kept out, by Native Americans who set frequent fires. The native grasses were brown in summer partly because there was seldom any rain after the winter. These grasses had deep roots, but not quite deep enough to reach the water table. European and Asian grasses, brought in later by the Spaniards, had shallow roots. The conquistadors saw a prairie landscape that was an unpromising brown during the growing season.

Right next to the dry prairie, marshes and open water formed Tulare Lake. This lake was the largest freshwater body entirely contained within what is now the United States, but it was shallow. Most of it was shallow enough that great mats of tule sedges (*Schoenoplectus acutus*) and tulare cattails (*Typha latifolia*) grew everywhere, except in the middle of the lake. The water came from Sierra snowmelt.

The first white explorers looked at a land that seemed almost worthless to them except for grazing cattle.[1] Half of the land was too dry, and half too wet, to support a large white population. In contrast, the Native Americans, especially of the Yokuts tribe, had a high population density and had figured out how to live prosperously off the land. They managed the grasslands with fire, obtaining food from them. They also ate fish and rabbits. Protein-rich acorns from the abundant oaks were also important (chapter 6). The conquerors, however, were not willing to live off of rabbits and acorns. During the nineteenth century, the whites nearly exterminated the Natives through murder and forced servitude, just as they did everywhere else in America. This was the blank page upon which white Americans would impose the strict order of agricultural utopia.

In the twentieth century, technology allowed groundwater to be pumped and snowmelt from the nearby mountains to be channeled into canals. Only then could the landscape be transformed into a garden that resembled a desert oasis more than it did an American farmland. Today, much of the groundwater has been pumped away, and agriculture depends largely on the canals. Nearly all the snowmelt that used to accumulate in Tulare

166   *Forgotten Landscapes*

Lake has been siphoned into agricultural and municipal use. All the heavy metals from nature and man, most notably selenium, have concentrated themselves in a squalid little pond that used to be the Kesterson National Wildlife Refuge.[2] Being so shallow, the lake area fluctuates, sometimes large after heavy winter rains, sometimes hardly existing at all. In the massive rains of 2023, the governor declared that Tulare Lake was back.[3]

Another reason that the San Joaquin Valley seemed at first like a bad place to create white civilization was disease. In the absence of reliable data, it is difficult to say whether it was any worse than anyplace else. It was not just malaria that the abundant mosquitoes from the lake carried, but also viral encephalitis, a brain infection.

When I was in elementary school, a local right-wing news broadcaster (back when they were rare) spread alarmist news about how we should be afraid, very afraid, of the abundant mosquitoes one summer after heavy winter rains. He said there was someone in a long-term care facility who was "a vegetable" because he got encephalitis from the bite of a *Culex tarsalis* mosquito. This was the first scientific name I ever learned. I dutifully wrote an article about mosquitoes and encephalitis for our elementary summer school newspaper, but our teacher removed the scary part and just inserted, "Make sure you get out your bug spray for the Fourth of July." Gee, thanks, Mrs. Webb, for gutting the first piece of popular science writing I ever published.

In addition, the county in which I grew up (Tulare County, named after the marsh plants) had a disease named after it, a credit that I think no other county can claim. Tularemia, also called rabbit fever, was common in the dry grasslands. I got it, felt bad, then recovered. By this point in the history of humankind, tularemia had evolved from an acute into a mild disease (chapter 7).

Unfortunately, even the canals did not provide enough water. During drought years, when there was little snowmelt, the San Joaquin Valley pumped massive amounts of groundwater. In 2000, long after I was gone, Fresno County immediately to our north led the nation with 3.7 billion gallons of groundwater pumping during a drought year.[4]

SHADOWS IN PARADISE

The San Joaquin Valley paradise, as seen through my childhood eyes, was illusory. Every paradise has its shadows, perhaps none more so than the San Joaquin Valley.[5]

Consequences of Pesticides

There was a price to pay for the kind of paradise that modern white agriculture and orchards have given to us. As I explained in chapters 4 and 6, the fields and orchards were monocultures of single crops, each of which is an unlimited feast for whatever insect pest might eat it. As a result, huge amounts of pesticide drenched the air and soil. Crop-duster biplanes sprayed massive amounts of pesticides all over the orchards and fields. Flimsy posters, attached to some of the citrus branches, proclaimed "Danger!" (*Peligro* in Spanish, for the Mexican farmworkers). When I grew up in the Valley, pesticides were the smell of the summer evening. To this day the scent of Malathion makes me homesick. One day my dad drove home from work right through a cloud of pesticide spray. He had to pull off the road before he could recover from the dizziness and nausea.

Pesticide toxicity is the same experience found in other places with irrigated production of fruits and vegetables for the American market. It is frequently the poor Mexican farmworkers who face the lethal health consequences of pesticide use.[6] The Valley has a culture that is very distinct from the rest of Southern California.[7] Part of its identity is living with the consequences of pesticide toxicity.[8]

In a way, the fatal dependence upon pesticides might be considered surprising, since one of the earliest attempts to use biological control on agricultural pests took place, with astonishing success over the long term, in the San Joaquin Valley. An outbreak of cottony cushion scale (*Icerya purchasi*) on citrus trees had reached epidemic proportions by 1888. Then agricultural researchers imported vedalia beetles (*Rodolia cardinalis*) from Australia. It is still considered one of the best examples of using a predatory insect to fight against a pest insect.[9] Because the predator populations increase along with the pest populations, the control sustains itself over the long term without pesticides. In fact, in some orange groves near where I grew up, there are now signs legally prohibiting the use of pesticides, though I did not see this until long after I had moved away.

The heavy use of pesticides not only resulted in toxic residues but also in the pesticides becoming useless. As was first pointed out to popular readers by Rachel Carson in 1962, the pesticides constitute a strong natural selection upon pest populations, favoring any pests who happen to be born with any of a number of ways of resisting the pesticide, whether by avoiding it or detoxifying it.[10]

At the time I grew up in the Valley, persistent pesticides such as DDT were still in use. These molecules can not only build up in the bodies of

168   *Forgotten Landscapes*

animals, but can build up through the food chain, reaching deadly concentrations in the fatty tissues of top predators. The use of persistent pesticides has been largely discontinued. Instead, mostly pesticides that degrade upon exposure to ultraviolet light and oxygen are used. Soil microbes can also break down some of the non-persistent pesticides. This is supposed to mean that pesticide residues break down before they have a chance to build up, but this occurs significantly only in the presence of light and oxygen. In many cases, pesticides either build up in the soil or wash away to other locations where they were not intended to be.[11]

Consequences of Irrigation

Because of the dry summers, irrigation is necessary. The vast majority of California water is used for irrigation, something that urban developers do not like, but which, if irrigation were done correctly, might be worth the cost. However, irrigation as it is currently practiced in most places cannot be sustained into the future, for reasons explained in chapter 5. This includes salt buildup.

As I noted in chapter 5, in California, water flows uphill toward money. The biggest farmers have huge political power in California, especially in the San Joaquin Valley. In some communities, such as Arvin, almost all of the farmland belongs to rich farmers. They generate little income for the local people, who are mostly poor farm laborers. In contrast, communities such as Dinuba have many small farms and a thriving middle class. It is the availability of water that makes land either very valuable, or not.[12]

In some places, the land itself has subsided by eight and a half meters since groundwater pumping began in the 1920s.[13] As practiced in the valley, irrigation is an unsustainable practice that probably would not have occurred if Native tribes rather than industrial white Americans had developed the San Joaquin Valley.

Air Quality

Los Angeles, with its smog, was a hundred miles to the south, on the other side of the Tehachapi Pass from the San Joaquin Valley. I was happy to be a country boy away from the polluted cities.

I did not realize, when I lived there, that the air pollution in the San Joaquin Valley was worse than that of Los Angeles or New York City—in fact, the worst in the United States. By almost any measure, the worst three places for air pollution in America are Bakersfield, Visalia, and Fresno—all

A Toxic Paradise 169

Figure 8.4. A temperature inversion layer that holds in the Valley air pollution is best visible at sunset from the Sierras to the east. *Author photo.*

in the San Joaquin Valley.[14] A temperature inversion layer held in the dust, pesticides, and hydrocarbons (figure 8.4), except when rain washed them out—which is why we could see the Sierras only briefly after winter rains.

Artificial Crops

Most crop plants are artificial in the sense that they have been bred to be different from their wild ancestors. In general, the Natives used fewer specially bred crop plants than do modern farmers, but it was the Natives of the Mexican highlands who bred an unpromising wild grass called teosinte into modern maize, which could not survive in the wild. The Natives of the valley, however, did not practice agriculture.

Many of the oranges grown in the Valley are seedless. These navel oranges are so called because a mutation produces a fruit with an undeveloped second fruit at the end, away from the branch, which reminded somebody of a human navel. This mutation also prevented seed production. In nature, seedlessness would be a guarantee of extinction. But in our artificial paradise, orchard trees do not need to produce seeds. We can propagate seedless orange trees by grafting. First, you grow a hardy orange tree until its root system is large enough. The hardy oranges I saw resembled what I now recognize as bergamot oranges, whose fruits are sour and filled with

170   *Forgotten Landscapes*

seeds. Then you clip the trunk away, leaving a little stump. To this stump you graft a branch from a navel orange tree. If done correctly, no pathogens infect the wound, which bleeds sugar from the phloem, and stem cells proliferate, forming connections between branch and trunk that allow the upward transport of water and downward transport of sugar. This is the same process used in establishing walnut orchards (chapter 6).

Artificial Warmth

The orange orchards do not even have to be perfectly adapted to the temperature conditions. While an orange tree can survive a mild frost, a deep freeze (even for a single night) can kill it. Or if not, the freeze can damage the flesh of the orange enough to make it unpalatable and unmarketable. On a cold winter night (that is, just barely below freezing), a degree or two of freezing down inside the fruit can spell the difference between making a million dollars and losing it. To keep oranges from freezing, agricultural researchers enlisted the laws of physics.

The pulp of the oranges is sweet, because they have been bred for sweetness. The sugar molecules act as a kind of antifreeze. This is one reason that the orange pulp, to be damaged, must be exposed to more than just a mild frost.

Often when I was in bed as a child in the Valley, I would wake up to hearing what sounded like airplane engines, as if a World War II–style invasion was under way. In fact, airplane engines were exactly what I was hearing, but they were not on airplanes. I was hearing wind machines, which are airplane propellers perched on top of poles above the level of the orange trees (figure 8.5). The orange growers use wind machines as an attempt to keep the oranges from freezing.

This is how it works: Imagine yourself camping out under the stars. Even if the air is warm, you can feel the chill. The reason is that your body is glowing with invisible, infrared light. Every object in the universe emits infrared light. The warmer the object is, the more infrared light it emits. When you are indoors, the walls and furniture and floor also emit this light. You are warm-blooded, and your body emits more of this light than the objects around you, but they partly compensate for your energy loss. If you camp under trees, the leaves emit infrared light, partially making up for the infrared that you have lost. If you camp under the clouds, the clouds emit their own invisible light.

But in outer space, there is almost no infrared light in the vast expanses between the stars. The temperature of outer space is almost absolute zero,

A Toxic Paradise   171

Figure 8.5. A "wind machine" is an airplane propellor on top of a pole. On cold clear winter nights, the propellor blades keep the air stirred up just enough to prevent the orange fruits from freezing. *Author photo.*

except for a little bit of infrared left over from the Big Bang. If you camp under a clear sky, your body loses infrared, and the sky provides none in return. The molecules of the atmosphere are not dense enough to produce a significant amount of their own infrared. You receive infrared from the ground, but not from the sky. That is why any object, such as yourself, that is out in the open under the stars can lose energy and become cooler than the surrounding air.

172   *Forgotten Landscapes*

On a cold clear night, the air near the ground can in this way become colder than the air above the trees. When this happens, the oranges can freeze even if the air above the trees is higher than the freezing point. That is, unless the wind is blowing. But in the Valley, the coldest nights are often utterly still. The secret is to get the air to move, even just a little. And that is just what the wind machines do. They stir up the air, bringing the slightly warm air above the trees right up next to the oranges. The wind machines are fans, but in this case, instead of blowing away the heat, you can think of them as blowing away the cold.

Artificial Ecology

In the springtime, I prowled among the ruins of an old prison camp that had been used to hold German soldiers during World War II. In this highly disturbed soil, I found many species of wildflowers, almost as many as on the surrounding hillsides. Some of them, such as the fiddlenecks, red maids, poppies, and gilias, were native species. Others, such as black mustard, had been deliberately introduced by the Spaniards centuries earlier.

Human activities, such as agriculture and cities, have disturbed almost all of the San Joaquin Valley. In one study of the plants of the valley floor, introduced weeds such as filaree (*Erodium cicutarium*) covered 63 percent of the disturbed area.[15] The grasses, including wild oats, were almost all Central Asian or European in origin, the native grasses having been driven almost to extinction, particularly by grazing animals to which they had never evolved, and by plowing.

To me it was a wonderland. I chased meadowlarks in these disturbed fields. I loved all of these places, but they were all artificial experiences.

The San Joaquin Valley is far from being the only artificial environment, of course. Many urban areas (and even worse, areas of suburban sprawl) are almost entirely artificial except for sneaky weeds that escape our control. And the whole science of forestry began as an attempt to replace native forests of diverse tree species with uniformly spaced trees of single fast-growing species. In a pine plantation, a dogwood tree is a weed.

The modern San Joaquin Valley is, in nearly every way, artificial. It is completely different not only from what it was before white contact, but from the way it would have been if Native Americans had developed it. The white approach to living in a place like the San Joaquin Valley is to erase as much nature as possible and create something that is artificial. In Native culture, and in nature, you almost never see straight lines, which is the dominant feature of my beloved toxic paradise, the San Joaquin Valley.

9

TOWARD A HEALTHIER WORLD, WITH A LITTLE HELP FROM YOUR LOCAL NATIVES

White-dominated America needs to pay attention to Native American experiences. Native cultures have insights that could help to save the world. They are not vague spiritualistic insights based on the use of Native herbs, either; they are concepts, such as habitat management and the promotion of biodiversity within agriculture, as explained in the preceding chapters, by which Natives positively transformed the American continent in the past. We did it before, and we can do it again.

To learn Native insights about people and the land, I would not suggest going to your nearest Native community and asking around. Many of these communities have been ground into poverty—and, in some cases, hopelessness—by centuries of deliberate oppression and many decades of economic hardship. Tribes vary greatly today in their level of health. For example, the Cherokee tribe, of which I am a member, has excellent health care and environmental services available. But it is the largest tribe in America and has a lot more money than some other tribes. Even for Cherokees, I would not suggest you drive around in the hills of northeastern Oklahoma and ask questions. They won't want you asking about them.

Many Natives have forgotten all about their past, except that it was grim. The commonly held image of Native Americans is that we are dirt-colored drunks passed out in the ditch on the Rez in flyover country. To the extent that this image is ever true at all, it is the product of white conquest, from which no tribe has escaped. Instead, what is necessary is what I have done in this book: examining the historical record to see what the Native tribes did back when their communities were still relatively intact. At the same time, you can always find small—or large—groups of individuals within each tribe who are reconnecting with their healthy past.

174   *Forgotten Landscapes*

A NATION OF WEEDS

Perhaps the biggest difference between original Native communities and the conquering white culture is the attitude toward Earth.

The idea that Americans today have the right to unlimited acquisition of resources has long been a source of irritation to international bodies who are trying to deal with the problems that have resulted from it. At the United Nations Conference on Environment and Development in Rio de Janeiro in 1992, President George H. W. Bush said, "The American way of life is not up for negotiations. Period." Many attendees took this to mean that Americans would never consider restricting their use of natural resources to any extent or for any reason. This was in the early days of international attention to global warming—the result of the production of carbon dioxide by human economic activities—and at the time, America was the leading producer of carbon emissions. Bush's message to the world appeared to be that the number-one source of the problem was not going to take any steps toward solving it. Period.

And just when you thought it couldn't get any worse, his son, President George W. Bush, said "Goodbye from the world's biggest polluter" as he left the G8 economic summit meeting in Japan in 2008. The second Donald Trump administration has maintained and intensified this message.[1] We Americans have developed this attitude toward the world largely because we grew as a nation of weeds. Weeds grow rapidly in abandoned land during their single year of life. They commandeer all the resources they can without having to share them, saving nothing back for the future, and producing an explosive crop of seeds. The seedlings then do the same the following generation, since unlimited resources are still available. Slower-growing plants such as goldenrods store back some of their wealth safely underground for the future. Bushes and trees store back some of their wealth not only underground but in aboveground stems, each with its own bud ready to grow again. Seedlings of forest trees grow slowly in the shade and wait for their chance to emerge into the sunlight. We Americans, like weeds, grew our country as if there was no future.

The problem is that the weedy way of life works only as long as the world around them is not crowded. That situation, however, cannot last. The planet and its resources are finite, although our economic system ignores this undeniable fact. A small population of bacteria, placed in a test tube of broth, can have a population explosion, each bacterium producing trillions of offspring, but only until the bacteria deplete the broth and fill it with their wastes. Weeds can have a population explosion in a spot of land

Toward a Healthier World, with a Little Help from Your Local Natives   175

where other plant species have been cleared away, but population explosions are always temporary. As economist Kenneth Boulding said, in effect, anyone who believes that unlimited growth is possible on a finite planet is either a madman or an economist.

This is exactly the reason that weeds dominate newly disturbed fields only for a brief time. Within a few years, they are displaced by other plants that can use the resources more efficiently and can invest in the future. People with a weedy attitude toward their world will be displaced, just like the weeds. Watch out, America; the future is not yours.

American colonists moved into a landscape that was almost unbelievably good and surprisingly empty of human habitation. They assumed that God had given them this good land for their own use. But it was not God who did it. It was the Native Americans, who had cleared out forest undergrowth and opened the land for cultivation—their own cultivation, until the European conquerors took it. The Natives transformed the landscape in the ways described in this book. The Natives got out of the way by either dying or by fleeing to the West, where they displaced other Native tribes. The Natives left behind them a resource space for unlimited white European population growth (chapter 7).

Many economists point out that technological breakthroughs can increase the resources available for economic growth. An example one hears a lot about today is Moore's Law, which says that computing power keeps doubling about once every two years. The fact that I am writing this book on a desktop computer that can process megabytes of digital information almost instantly, instead of using punch cards and a typewriter, the way I did a half-century ago, seems to confirm Moore's Law. But computing power is limited, even if we have not yet reached that limit. At the very least, a bit of computer information cannot be less than one atom. In fact, a bit would require considerably more than one atom because of quantum fluctuations in atomic states. So you cannot take comfort and imagine that Earth's food resources will keep doubling forever any more than you can imagine that computing power will double forever.

History is full of examples of temporary population explosions following technological advance. When the Eastern tribes pushed the Dakota from the forests out into the prairies, the Dakota were not able to make a very good living out there. That is, until they got horses. Horses had become extinct in North America at the end of the most recent ice age. The horses that the Dakota encountered were descendants of animals brought to North America by the Spaniards. It didn't take long at all for the Dakota to learn to ride on, hunt from, and fight from horseback. It took scarcely a

176   *Forgotten Landscapes*

century for them to become some of the best horsemen in the world. Very quickly, their food resources increased because they no longer had to sneak up on bison; they could chase them down.

The problem with technological breakthroughs is that you cannot count on them. Suppose the Spaniards had not brought horses. The explosion in the Dakota resource base would not have happened. Another problem is that it is one thing to think of breakthroughs, and another thing to develop them. Are we, as many scholars think, running out of fossil fuels? We can always use solar power. But this option is not available without deliberate investment in solar technology.

These are the circumstances under which the American mind evolved: Economic growth was always possible because there were always more resources. Kenneth Boulding called this the "cowboy economy." From our pioneer days, we have hated the idea of limits. Even when we intellectually admit that the world is limited, we refuse to accept it emotionally. We think that an economy that grows *only* 2 percent in a year is essentially a failure. This started as a largely white American view. The Native Americans, in contrast, were perfectly happy with a society that was the same size this year as it was in the past.

The Europeans who stayed in Europe learned to live within limits that white Americans have never recognized. Even I, who use what I consider to be a small amount of energy and resources, remain surprised at my new French neighbors, who often consider air-conditioning to be an unnecessary luxury.

The attitude of unlimited consumption can also be seen in the food industry. Food corporations spend billions of dollars on advertising, and on product placement, to convince us consumers that we can eat anything, anytime, and as much as we want—or, as much as their advertising makes us desire. The result is sickness associated with a grossly excessive diet. Conveniently, there are other corporations that can sell us physical and psychological ways to solve the resulting health problems, unless we die first. And, of course, dying is expensive, if you do it in the gleaming-casket American way.

Perhaps the white attitude of unlimited greed is best expressed by the Cherokee warrior Tsiyu Gansini, in the musical *Nanyehi* (based on the life of Nancy Ward) by Becky Hobbs and Nick Sweet:

> *You have a hunger*
> *That will not die*
> *You must take everything*
> *I don't know why.*

Toward a Healthier World, with a Little Help from Your Local Natives   177

It is not useful to assign blame. The only useful thing to do is to ask if there are any insights or beliefs that the Native cultures did in fact have—and that the Europeans and white Americans did not—that can help to bring the world back from the brink. As we make progress on the unfinished work of rediscovering Native cultures, are there some things the larger white society should adopt from those cultures? The answer is yes, and I have given examples throughout this book.

Agroforestry

One of these examples is agroforestry (chapter 4). Our modern industrial agriculture produces a lot of food, but at the cost of soil erosion, exploitation of water resources, and toxic chemicals. The fact that many people do not have enough to eat is not the fault of our agricultural system, but rather of what we choose to plant and raise.

A large proportion of American agricultural production is not food for humans but for livestock. Americans eat more meat than anyone else in the world, and this takes up a lot of farmland for corn and soybeans to feed the livestock. We raise this animal feed in large monocultures that generate a lot of soil erosion and toxic runoff. We are particularly proud that one American farmer can produce, through chemicals and machinery, enough food to maintain at least a hundred other people. So productive is our industrial food system that the government often has to pay farmers to not produce as much as they could.

Native polyculture and agroforestry could not have, under any circumstances, produced this much agricultural output. It would take a lot of digging sticks to produce corn for the huge feedlots that now dominate large areas of the rural landscape. But the Native attitude was to be content with less. When Elizabeth's great-grandmother Nancy Ward introduced beef cattle to the Cherokee economy somewhere around 1800, she could not have imagined the huge feedlots we would have today.

An agricultural idea closer to the Native mind-set would be the small, sustainable farms that raise "organic" produce (meaning, without fertilizers or pesticides), almost always more than one kind of crop, and then sell them at local farmers' markets. While organic farms cannot produce as much food as industrial monocultures, they can produce enough to shift the market in a healthy direction. When I go to our local farmers' markets, formerly in Oklahoma and now in France, I see a lot of people buying a lot of food, and these are all people who are not buying those foods, at least, at supermarkets. The customers make selections carefully, thinking about eating healthy foods.

178   *Forgotten Landscapes*

During the time of World War II food rationing—an experience we can no longer even imagine—the government encouraged Americans to buy only what they needed, to cook it carefully, and to eat all of it. Every consumer had to have ration coupons to buy things like sugar and meat. My Cherokee grandfather not only lived within these limits, but by the end of the war he had a bunch of now-useless ration coupons left over. He was a farmer, but grew meat and vegetables for his family, not for market, except for the cream which he hauled down to the store to sell. Our planet would benefit, as would our health, if we bought less, and healthier, food. My grandfather did not think of himself as an environmentalist, but he lived like one.

A second example is that our destructive economic system often leads to species extinction. Natives never thought in terms of destroying a habitat. When they set fires, they were confident that the trees or grasses would grow back, and that they were doing something that would benefit wildlife in the area. Even if they could have imagined the possibility, it is unlikely that the Hohokam cities would have dammed an entire canyon to irrigate their crops.

Native attention to biodiversity was a matter of survival. To a much greater extent than in our modern white economy, even agricultural Natives depended on wild plants and animals. Brain studies have shown that Asians (and I think this applies to Native Americans also) tend to notice the environment more in illustrations than do cultural whites, who notice almost exclusively the person or animal in the center of the illustration. We need to start re-seeing the rest of the world, not just ourselves at its center.[2]

Global Climate Change

Perhaps the major problem facing the modern world today and into the future is global climate change. One component of climate change is global warming. In even the recent past, the problem was not so much the heat itself but the associated problems, such as drought. To a certain extent this is still true: You don't see people falling over dead from global warming. But it will not be long before large portions of Earth become uninhabitable due to heat.

Already, during the summer 2023 heat wave in America, roads began to crack and buckle, interrupting transportation. If heat reduces our transportation, and melting glaciers raise the ocean level to flood billions of dollars' worth of coastal land, and drought destroys crops, it hardly matters if the heat is not high enough to kill you. Stronger storms, associated with global climate change, may not have a globally noticeable death rate, but they are causing insurance companies to pull out of some high-risk areas such as California and Florida, where storms and fires are even now causing major infrastructure losses.[3] Global warming—we can't afford it.

Global climate change has resulted from our uninterrupted production of greenhouse gases such as carbon dioxide as a result of industry, agriculture, and our personal energy use. As the world population increases, we all need to reduce our carbon footprint, which results not only from our daily choices of transportation and food, but also the industry that provides the fuel and food for us. The solution is so simple as to be overwhelming. We just need to use less energy and materials. This is a Native attitude. Plants, including trees, remove carbon dioxide from the air in order to make their own food by the process of photosynthesis. We also need to adopt the Native attitude of letting the trees grow rather than cutting them down.[4]

Not only that, but traditional Native American societies were resilient in the face of natural climate change. After arriving in North America, they experienced not one but two episodes of dramatic climate change.

The first episode was the global warming that occurred at the end of the most recent ice age, about fifteen thousand years ago. As the first tribes migrated south from Siberia, they saw the glaciers melting and trees growing where tundra had once been. They saw the sea level rise by as much as three hundred feet. No single individual may have noticed these changes, although an individual could have seen a glacier retreat during his or her lifetime had he or she stayed in one place and thought to notice. Did their wild food plants or prey animals die out in one location? They could either learn to hunt and gather something else, or move someplace else to find their old foods, if they knew where to go.

This option is no longer available. When sea levels rose, the Natives just moved inland, leaving behind archaeological sites, some of which are now below sea level. Americans cannot do this today. If you move inland, you have lost the value of your coastal home and somebody already owns "inland." Follow the migration of the game to a new location, the way the Natives did? But today the game cannot move, since there are cities, highways, and farms in the way.

The second climate change episode Natives faced was that, after the global warming had begun, about thirteen thousand years ago, the warming reversed, and glaciers returned temporarily for a couple of centuries. This period is called the Younger Dryas, because pollen of the arctic plant *Dryas* reappeared in glacial melt sediments after they had previously disappeared due to the warming climate. Natives adjusted to this also.

180 *Forgotten Landscapes*

Part of the reason that Native tribes survived the global warming and temporary global cooling at the end of the last ice age is that their populations were very low. Once they built up larger populations, by about a thousand years ago, they became more dependent on agriculture. They couldn't just pick up and move. They were then more vulnerable to climate shifts.[5] It is likely that the small populations of Natives, say, ten thousand years ago, could have easily survived the drought that contributed to the collapse of the Mississippian culture (chapter 1) and to the desert southwest cultures (chapter 5).

Even after the drought eight hundred years ago, Natives reached a new equilibrium very quickly. Their cities had collapsed, but their farms and trade networks continued in a society consisting of large interconnected villages rather than cities (chapter 1). They did this because they were more willing to live within the limits of their environments.

The common theme to these examples is that the white viewpoint has been and continues to be to acquire as much as possible, while the Native viewpoint has been to be satisfied with having enough. To take a Native attitude by using less energy and less stuff—this is the low-hanging fruit to combat climate change. It is not a technological solution, nor would laws based on it be enforceable. But, many of us are convinced, nothing else will work. We need to be more like the trees of an old forest, and less like weeds. Period.

CULTURAL DIVERSITY

Another way in which the Native attitude was superior to the white culture that replaced it is its recognition and appreciation of cultural diversity.

A society pervaded by Native American ideas would almost certainly have more appreciation of cultural diversity. Cultural uniformity dominates white culture; in particular, the cultural elements that major corporations can convince us to buy. Everybody in our culture recognizes both hard and soft drinks, and even tea, by their brand names. If you want to drink water, you should drink it from a container that identifies it as being bottled thousands of miles away on an island in the Pacific Ocean. And if you want to drink horchata? This is marketed only to Hispanics.

In contrast, Native Americans never lived in a culturally uniform environment. No tribe was big enough that a person could be sure of never encountering someone from another tribe. There are hundreds of Native languages, many so different from one another as to be mutually unintelligible. A Cherokee talking to a Navajo? No wonder Native American tribes relied extensively on sign language. Although hostile tribes could be cruel to one another, captured members of another tribe were often adopted into the tribe of their captors. It was usually the women who got to decide the fate of a captive. Living in an environment of cultural diversity is true of tribal peoples all over the world. According to Jared Diamond, it is not unusual for rural New Guineans to speak six languages.[6]

NATIVE INSIGHTS

This book assembles evidence that Native Americans have always been, since their arrival in North America and after their populations built up significantly, an important force in the life of the landscape. All I am asking, in the name of my Cherokee ancestors and fellow Natives, is that our importance in the natural history of the continent be recognized. It is part of my responsibility as a scientist (plant ecologist), science writer, and science educator to champion this viewpoint. Just as I teach and write about natural ecosystems and regenerative agriculture, I also teach and write about the role that Native Americans have had in them. We have lessons to learn today in modern America, about how to grow our food and manage our forests, from the Native traditions.

I do not mean to imply that pre-contact Native cultures were some kind of golden age. North America may have been the best place in the world to live, but it was not perfect. Life was difficult, suffering was continuous, and there was violence and oppression—in Native cultures, just as in all other cultures in the world. Humans are humans. But many practices introduced by the white culture destroyed more than they helped. At the very least, the white culture was not superior, and did not have the right to label Native cultures as savage as a justification for eliminating or erasing them. White culture imposed a European and white American mind-set on a landscape that the Natives had learned, over many millennia, to nudge in the direction of health and productivity.

Nor do I mean to imply that we should go completely back to the way the Natives did things when they were in charge of the continent. Even the practices that were healthy for the North American continent when the Natives possessed it might now be injurious.

Fire is one example. Burning a forest or grassland renewed it back in Native days. Today, a controlled burn does indeed encourage the growth of grasses, but not necessarily native grasses. One such burn in a forest near

182   *Forgotten Landscapes*

Tulsa produced a luxuriant growth of Johnson grass (*Sorghum halepense*), a grass that came from Asia and has been displacing American native grass species. Practically all the grass on eastern Oklahoma roadsides is Johnson grass, recognizable by the bright white midribs of the leaves. Burning only gives Johnson grass a yet greater advantage over the native species.

In the drier grasslands of the West, European and Asian grass species such as wild oats (*Avena barbata*) and cheatgrass (*Bromus tectorum*) have displaced native grasses from millions of acres. Other invasive species have taken permanent hold. We will now always have Asian puncturevine (*Tribulus terrestris*) on dry roadsides and the Asian tree of heaven (*Ailanthus altissima*) at the edges of our forests. California will always have eucalyptus trees, which white conquerors introduced from Australia as a quick-growing source of wood. The native bacteria and invertebrates cannot break down the toxic eucalyptus leaves, which poison any native plants that might happen to start growing underneath them.

Even if we succeeded in reinstituting a healthy system of fire management, without first addressing the problem of invasive species, these species might benefit from it more than the native plant species. And if we bring global warming under control, we will still have the invasive species pushing their way into our natural habitats.

If you take a Native viewpoint, you can probably figure out what you can do to make the world healthier. You probably don't need to hire a Native consultant. You probably don't need any sacred Cherokee herbs. Just pretend you are wearing moccasins and walking through the world. I think the dominant white culture can learn a lot from ancient Native practices. No matter how many technological innovations we have to counteract the damage our white culture has inflicted upon the Earth, nothing will matter unless we re-see the world through traditional Native eyes. There are many cultures, Native American and others, from which we can pick and choose the best insights.

If we do this, our world will be a little bit less of the Darkening Land that Elizabeth feared it would become.

EPILOGUE

What We Have Lost, and What We Can Regain

When Europeans first sailed to North America, they thought they had found a wilderness. Native Americans met them at the boat, but the immigrants dismissed them as savages who hardly deserved to keep the land that God had so richly blessed, believing the Natives hardly used the wilderness that God had opened up for the immigrants to conquer and transform.

The immigrants were only too aware of the mess they had made of the Old World and thought they could start over again in what they called the New World, with a clean slate on which to write their myth of holiness. They would get it right this time—a continent on which they could establish empires that were perfect utopias of godliness.

Though most Americans will not admit it, this is a fundamental component of the white American mythology. A mythology consists of big-picture stories that give us meaning, individually and as societies. It may be built on true stories, or false ones, but it hardly makes any difference: It defines who we are. One expression of the white American mythology is in the words of the seldom-noticed second verse of "America the Beautiful" by Katharine Lee Bates, sung by earlier generations of schoolchildren, even the ones whose ancestors were conquered or enslaved by the whites:

> *O beautiful for pilgrim feet,*
> *Whose stern impassioned stress*
> *A thoroughfare for freedom beat*
> *Across the wilderness!*

The settlers believed that the trees grew and the streams flowed the same as if the Natives were not present. If the Natives did have an impact

184   *Forgotten Landscapes*

on the land, it was only in the immediate vicinity of their villages, where their tepees were clustered around a campfire in a small barren area. As Caliban-like children of the forest, the Natives were a little bit scary. They will come and steal *your* children if you give them half a chance.[1]

In the centuries since the Europeans arrived, most Americans have come to believe that the conquest of North America was often cruel by modern standards, but justified, because North America was a wilderness. This was not true. The New World was already somebody's home; not their campground, but their thickly settled, thoroughly transformed home. Invading America was no more justified than invading another European country.

Many of the immigrants thought that, rather than conquering the land, they were awakening it to its possibilities as a productive landscape. This was the idea behind Conrad Richter's trilogy *The Awakening Land*. But Natives had already transformed North America into a productive landscape. The immigrants just refused to see how the Natives had already awakened it. At the end of Richter's trilogy, the main character, Sayward Luckett, now an old woman, lies in her bed in the Ohio city that had completely replaced the forest during her lifetime. Through her window she sees one tree that escaped the ax and begins to wonder if the land had really been awakened, or if it had been destroyed.

When Daniel Boone led settlers over the Appalachian Mountains into Cherokee land, he and his people thought it was as pristine as the day God made it.[2] Crystal-clear streams, rich soil, tall trees, green grasslands, just waiting to be turned into cities of hope and joy, or at least profitability. To some of the settlers, "taming the wilderness" was literally the will of God. They believed that white settlers moving onto Indian land was just like the Israelites entering the promised land of Canaan. There was a popular religious song at the beginning of the nineteenth century in Kentucky which literally called it the Promised Land. Entitled "On Jordan's Stormy Banks I Stand," it was written in 1787 by Samuel Stennett. Probably some of Boone's crew were singing this song as they walked and rode over the dirt trails.

> *On Jordan's stormy banks I stand*
> *And cast a wishful eye*
> *To Canaan's fair and happy land*
> *Where my possessions lie.*

The settlers were aware that Native Americans—still called Indians because of Columbus's confusion about where he had landed—lived in the hills and plains. The Old Testament referred to Canaan as a land flowing with milk and honey. But the milk came from cattle and the honey from beehives maintained by Canaanites, who had to be killed. This was the same thing the settlers believed about the American wilderness, although if the Natives would just get out of the way, it would not be necessary to kill them.

There were also paintings from the nineteenth century which showed white conquest of the American landscape as if the Natives were at best an impediment. For example, the 1872 painting *American Progress* by John Gast shows white pioneers and farmers, along with railroads, moving westward while bison, wolves, and Natives fled in terror, away from the bright white dawn in the east and toward a thundering darkness in the west. A voluptuous angel, who is Columbia, the spirit of America, wears revealing flimsy white raiment as she leads them. This is the version of American history that was taught in schools at that time, and long into the twentieth century. The angel in the original painting holds a book labeled "School Book" (see figure below).

The consequences for understanding the white conquest of America could not be greater. It is the difference between squatting on somebody's

Figure E.1. The 1872 painting "American Progress" by John Gast is currently in the Autry Museum of the American West. *Wikimedia Commons.*

186   *Forgotten Landscapes*

tract of hunting land, an act that is wrong but not outrageous, and barging into someone's home and driving them out, burning down the house and calling the land your own, an act that anyone would recognize as evil. The consequences are also great for our modern understanding of the plight of Native Americans today, largely reduced to poverty and powerlessness.

ERASURE OF NATIVE CULTURE

Genocide

The deliberate erasure of Native culture began as soon as Columbus set foot in America, and has continued until the present time. Even the most basic question—How many Native Americans were there in North America before European contact?—has been a political rather than an academic issue. If there were only a few hundred thousand Natives in North America prior to European contact, then the disappearance of Native cultures was more of a displacement than a genocide. Once the full extent of Native depopulation, directly or indirectly at the hands of the Europeans (and, later, Americans), was revealed, the enormity of the conquest could no longer be ignored.[3]

As I explained in chapter 7, most Native deaths resulted from exposure to Old World diseases. But there were also a lot of direct slaughters. Most of these remain buried in the annals of American history, largely unread by even the most educated general readers. For example, we think of English colonists as being religious pilgrims who came to America for freedom. But that is not really what they wanted. They left England for religious freedom, and went to Holland, where they found it. What they did not find in Holland was the freedom to impose their beliefs on other people. They had their chance to do this when they sailed to Massachusetts. Once there, they learned agriculture from the Natives, including the Pequot tribe. They repaid the generosity of the Pequot by slaughtering them.

Early in the seventeenth century, near Mystic, Connecticut, English soldiers surrounded a Pequot village at dawn and set it on fire. Then, when the people were running out, the soldiers shot them. All of them, men, women, and children. They trapped as many of the Indians as they could inside and listened to them burn to death. They could smell the burning flesh. One of the generals later declared that the attack on the Pequot village was the act of a God who "laughed his Enemies and the Enemies of his People to scorn." He boasted that he had made the Pequot village into

Epilogue   187

"a fiery oven." "Thus did the Lord judge among the Heathen," he wrote. Of the five hundred Pequots in the village, only seven were taken prisoner, and another seven escaped into the forest. There were hundreds of other similar massacres across the continent.

Ignoring Native Accomplishments

Most of the erasure of Native culture has been achieved by ignoring it, often deliberately. One good example of this is the book *A Nation of Immigrants* by John F. Kennedy, published posthumously.[4] Aside from one small paragraph near the beginning—in contrast to several paragraphs about Alexis de Tocqueville writing about Europeans who settled America—Kennedy completely ignored the Natives who were already here when successive waves of immigrants arrived. And even then, Kennedy claimed that the "Indians" were themselves immigrants who had displaced the "aborigines" who were already here—something for which no evidence exists, either today or at the time Kennedy wrote the book (which didn't stop him from reiterating this claim in chapter 5 of his book).

Most Americans today know almost nothing about Native history. How many people have even heard about Mississippian culture? How many know about the plagues of smallpox and other contagious diseases that nearly eliminated entire Native tribes? How many people have heard about the direct slaughter of Native warriors and civilians? How many people know that Native Americans were enslaved by Europeans and even constituted a significant portion of the European slave market, before so many Natives died that Europeans, and later Americans, had to rely on Black slavery?

Nearly everyone has heard about the Boston Massacre, in which five American colonists were killed by British soldiers, a number orders of magnitude less than the Natives killed by American soldiers and by the cowboys who have been so lionized in popular American culture.

Some of the erasure went beyond merely ignoring Native accomplishments. Sometimes it was erased deliberately, as in the bulldozing of Spiro Mounds (chapter 1) or of Hohokam irrigation canals (chapter 5). Another example is the forced reeducation of Native children in the boarding schools, in which the children were forbidden to practice Native culture or to even speak their Native languages. Well known in fiction, including *The Ordeal of Running Standing* by Thomas Fall (1970) and *Stealing* by Margaret Verble (2023), these schools have become notorious because of the deliberate neglect of Native children and their resulting high death rate.[5]

188 Forgotten Landscapes

The deliberate neglect of Native cultures continues today, as indicated by a notorious statement by former Republican senator Rick Santorum in 2021. He is as different from Jack Kennedy as anyone could be, but he agreed with Kennedy on one point. He said, referring to immigrants of European origin,

> We came here and created a blank slate. We birthed a nation from nothing. I mean, there was nothing here. I mean, yes we have Native Americans, but candidly there isn't much Native American culture in American culture.[6]

One might assume that at least by the twentieth century, white cultures were no longer attempting to kill Native Americans. But that does not mean they did not fantasize about it. In the early twentieth century, Oklahoma state senator E. M. Landrum said,

> It is unfortunate so large a proportion of the state's population is of Indian blood. I almost wish that the Indians could be wiped off the earth since they have been a bone of contention since Columbus landed.[7]

Even those who promoted themselves as defenders of Native Americans held views that we today recognize as openly racist, as well as historically inaccurate. One of these men was Henry Cabot Lodge, who said in 1913,

> Neither in war nor in peace has the Indian been able to stand against or beside [the white man]. Sentimentalists have inveighed against the whites for this; but history teaches that inferior people must yield to a superior civilization in one way or another. . . . In most cases, the Indian kept faith when dealt with fairly, even when being gradually pushed backward from his hunting grounds. But at best he was a dirty savage, dwelling in squalor and filth and content therewith. In consequence, epidemic diseases have often decimated the tribes.[8]

Lodge was either extremely ignorant or else ignored the fact that the "squalor" was partly the result of the epidemic diseases and conquest.

The assault on Native Americans continues with the second Trump administration, which in its first week questioned whether Native Americans could even be considered American citizens.[9] Whether and to what extent this new policy might be implemented remains, as this book goes to press, unclear.

Epilogue 189

Land Allotment and Legal Maneuverings

One of the most formidable examples of the erasure of Native culture was the practice of land allotment. By the time of European contact, after the Mississippian and other oppressive Native governments had collapsed, the Native tribes held their land communally, rather than as private property. Whites considered this in itself to be a threat to European capitalism.

Probably every tribe experienced what the Cherokee tribe did, including my grandfather Edd, a descendant of Elizabeth Hilderbrand Pettit. When Edd was born in 1879 in the Cherokee portion of Indian Territory, the land belonged to the whole tribe. In 1904, he and all the others had to register for land allotment. By the time Edd died in 1959, nearly all memory of communal land ownership had vanished from the Cherokee tribe.

Already by 1883 the men who were to engineer the divide-and-conquer strategy against the Cherokee and other Oklahoma Native nations were meeting in the East to make their plans. Senator Henry L. Dawes, ultimately the author of the law that divided up Native land, and others took a journey to look at land in the Indian Territory. Dawes said that "the head chief" told them, "There was not a pauper in the nation and the nation did not owe a dollar. It built its own capitol . . . and it built its schools and its hospitals." Dawes interpreted this negatively. "Yet the defect in the system was apparent. They have got as far as they can go because they own their land in common . . . there is no enterprise to make your home any better than that of your neighbors. *There is no selfishness, which is at the bottom of civilization*" (emphasis mine).[10]

But American governments (at the federal, state, and county levels) added insult to injury. Once all the land allotments were assigned to tribal members and their families, there was a lot of Native land left over. The American governments then sold or gave this "surplus" land to whites so that by the time of Oklahoma statehood in 1907, Indian territory was more white than Native.

It was only after forcing the Natives to own individual allotments of property that the white governments discovered their mistake. The supposedly worthless land they gave to the Muskogee tribe in Oklahoma contained more oil than anywhere else in the world (at least, that was known at the time). The rich white businessmen wanted that oil so badly they could taste the blood of the Natives from whom they would take it, in any way they could.

One of the ways that the whites could get the oil was by getting themselves appointed, by the courts, to be guardians of Native orphan

190   *Forgotten Landscapes*

children who had allotments with oil resources. The white guardians then had complete control over the profits from the oil, until the orphans reached adulthood. The guardians had no accountability for what they did. In many cases, the courts appointed white guardians for *adult* Natives whom they considered (or pretended) were mentally defective. The guardians proudly described themselves as grafters. In one case, some Native children in the Muskogee tribe were the legal owners of an oil fortune, but they didn't know it. They were living in a tree and drinking water from a creek. Their guardian used their money until the crusading reformer Kate Barnard found out about it.[11]

The abuse of Native adults by white guardians was the idea behind the 1949 movie *Tulsa* ("the lusty, brawling saga of a city of adventure"), starring Susan Hayward and Robert Preston, Pedro Armendáriz as the oppressed Native man, and Chill Wills as a Will Rogers knockoff. It ended with a big fire as the Native man touched his cigarette to petroleum wastes that had spilled on his land, causing all the oil equipment to burn up. Most viewers recognized this movie as a damning indictment of the grafters, but the actors in the movie were warmly welcomed in Tulsa, which seemed unaware that notoriety was not the same as fame.

In the case of the Osage tribe, the treaty had included tribal mineral rights—perhaps unnoticed by white overseers—which included oil. The whites still wanted the oil, but in this case the only way to get it was by murder, as described in the book *Killers of the Flower Moon* and the award-winning 2023 Martin Scorsese movie based upon it.[12]

Dumping on Native Tribes

The Muskogee and Osage tribes had land with the most oil, but the Quapaw tribe, which has a little parcel of land in extreme northeastern Oklahoma, had big deposits of lead and zinc near the town of Picher, Oklahoma. Today this town is out in the middle of nowhere, but during World War I it was the important location of one of the largest deposits of lead and zinc in the world.

Half of the lead and zinc that the United States used in the war came from this one deposit, and mining continued after the Great War. Over $20 billion worth of ore was extracted between 1917 and 1947. At the peak of activity, 14,000 miners worked at the mine, and another 4,000 worked in support services. Many of the miners commuted to Picher by a trolley system that reached as far as Joplin and Carthage in Missouri. Eventually the high-quality ore was depleted, and mining operations ceased in 1967. The corporation left huge piles of slag, or chat, which is rock that does not have commercially useful amounts of metal but remains toxic.

There were fortunes to be made in Picher—but who actually made them? The mining corporation executives made a lot of money, and the miners got salaries. The people in the surrounding area, mostly Quapaw Natives, experienced the impact of environmental contamination. The corporation had left seventy million tons of mine tailings and thirty-six million tons of mill sand and sludge. The wind picked up contaminated dust, and contaminated water leached metal ions from the chat piles.

The mine tailings could still be used as gravel—that is, if you ignored the contamination. So even after the mine closed, a quarry operated, and people still lived in Picher. The federal government recognized the area as so contaminated that it was designated part of the Tar Creek Superfund Site in 1983. In 1994, the Indian Health Service sampled blood from Native children in the Picher vicinity and found that 35 percent of them had blood lead levels that the Centers for Disease Control considered toxic. The Environmental Protection Agency sampled surface soils from areas of high public contact, including schoolyards and day-care centers; these soils were also highly contaminated. People started moving away, and the school closed in 2009.

The piles of slag (see figure E.2 on next page) are eroding into an eerie sort of phantom cliffs. A few defiant Quapaws remain today, flying their tribal flag. The mining corporation solved the problem of toxic wastes the way many other corporations have done: Just dump it on the Indians.

ELIZABETH'S EXILE

If you had seen Elizabeth Hilderbrand Pettit when she started out on the Trail of Tears, you might not have even known she was Cherokee. Like nearly everyone else in the tribe, she wore clothes similar to those of white Americans. In fact, she was seven-eighths white. Her only Cherokee blood came from her great-grandmother Nancy Ward, born Nanyehi, the last *ghigau* (or Beloved Woman) of the Cherokee tribe.

Nancy Ward was the full-blooded leader who tried, without much success, to chart a tribal future free from constant war with the white invaders. Elizabeth could have stayed back in Tennessee and pretended to be white, but Cherokee is something you *are*, not something you look like. Elizabeth's home and all her fortune were in the Cherokee Nation, and the US government knew it. She stayed with her tribe on the Trail.

Figure E.2. What appear to be natural cliffs are actually eroding toxic slag heaps left by a mining corporation in Picher, Oklahoma. *Author photo.*

Another Cherokee on the Trail who could have passed for white was the most famous chief in Cherokee history, John Ross, the principal chief of the tribe at that time. Like Elizabeth, he was only one-eighth Cherokee, and didn't even speak the tribal language. He had to rely on other Cherokees such as Reverend Jesse Bushyhead to translate his speeches for the people. Most of my Cherokee physical ancestry came not from Elizabeth but from the full-blooded Ooskuhneh, who married Elizabeth's daughter Minerva.

The white American excuse for driving out the Cherokees was not valid for Elizabeth, or for anyone she knew. The Cherokee Nation was densely populated, probably as densely populated as it would be for a long time after the white Americans took the land. When Congress passed the Indian Removal Act in 1830, declaring that the Cherokee—along with at least seventeen other tribes, including the Chickasaw, Choctaw, and Muskogee—would be relocated to the West, some of the bill's defenders claimed that the Cherokees were savages who ate "vile reptiles." Remembering her last roast beef dinner before she was driven from her home, Elizabeth was really irked by that statement. She also remembered the fine European peach trees whose fruit she and her neighbors enjoyed—from their orchards, not from wild forests.

The Indian Removal Act barely passed—101 to 97 in the House, 28 to 19 in the Senate—because a sizable minority of lawmakers could see through the false reasoning used to justify it. It was nothing more than a white land grab, in which President Andrew Jackson may have had a strong financial interest.[13]

Elizabeth knew that, in addition to having productive agriculture in the eighteenth century, the Cherokees began to adopt a white American agricultural economy in the nineteenth. By the time of the 1824 tribal census, the Cherokee Nation had a population of 16,060. They had 18 schools, 36 gristmills, 13 sawmills, and 62 blacksmith shops (see table below). The Cherokee Nation had 46,732 pigs—although I am uncertain how they were counted for this census. I mean, one can count people and schools and even plows. But pigs?

Cherokee farmers would let the pigs go out in the woods during the day to forage for acorns, and then at night the pigs would come back to eat fallen peaches from the orchards while the owners drank peach brandy. Did one count the pigs in the day or in the evening? Could one assume each man knew the exact number of pigs he had? The point of this table is simply that the Cherokee Nation was not made up of savages at the time the white governments were trying to depict them as such. They even had

194   *Forgotten Landscapes*

Table E.1.   From the 1824 census of the Cherokee Nation, before the Trail of Tears

Cherokee population (part- and full-blooded)	16,060
Black enslaved population	1,277
White population living in the Nation	215
Schools	18
Students	314
Gristmills	36
Sawmills	13
Looms	762
Spinning wheels	2,486
Wagons	172
Plows	2,923
Horses	7,683
Cattle	22,531
Hogs	46,732
Sheep	2,566
Goats	430
Blacksmith shops	62
General stores	9

slaves, just like white society. This was the world on which the US Army descended in 1838 to start the Cherokees on the Trail of Tears.[14]

Even though there is no direct evidence that Elizabeth could read or write, she knew very well that just a few years previously, a Cherokee scholar named Sequoyah had almost single-handedly invented a writing system for the Cherokee language. She was probably aware that this accomplishment was unique among Native tribes. She could not have guessed that, by the end of her century, the Cherokee Nation would have the highest literacy rate in the world.[15]

Racial Identity

In 1829 the Cherokee legislature passed a bill outlawing the traditional practice of bigamy, in which white traders and soldiers would have a white wife at home, outside of tribal territory, but also have a "squaw wife" in the Nation. Americans did not consider it bigamy, since marriages between whites and Cherokees were not recognized by the United States. The white traders and military men liked having squaw wives, since it created good trading relationships with the tribe. And for the Cherokee women, it was a source of stability in times of economic or political upheaval.

Elizabeth herself came from a long line of white forefathers. The Irishman Bryant Ward married Elizabeth's full-blooded great-grand-

Epilogue 195

mother Nanyehi (who became Nancy Ward), even though he had a white wife in South Carolina. Nancy's daughter Betsy married General Joseph Martin, even though he had a white wife in Virginia. This turned out to be useful for Nancy and Betsy, who took refuge with Martin during a time of political unrest in the tribe. This political strife was partly caused by Martin himself, who threatened the Cherokees in a 1781 letter but then magnanimously offered peace if the Cherokees would surrender more of their land.

Elizabeth's mother Nannie married one of the German Hilderbrand brothers who ran a gristmill and ferry in Tennessee. Elizabeth herself had married James Pettit, an army officer who already had a white wife who lived outside of the Nation. Some novels, such as E. Sterling King's *The Wild Rose of Cherokee, or, Nancy Ward, the Pocahontas of the West* (1895), and Sam J. Slade's *As Long as the Rivers Run* (1972), showed white men romantically and sexually involved with Cherokee women, but in these two novels, the women (openly modeled after Nancy Ward) were half-white, as if full-blooded Native women (such as the historical Nancy Ward) were unacceptable as romantic partners in white society.

On the other hand, it was scandalous even among the most broadminded white societies for a white woman to marry a Cherokee man. When the white-educated Cherokee scholar Buck Waite (who changed his name to Elias Boudinot) married Connecticut socialite Harriet Gold, the white neighbors burned Boudinot and Gold in effigy in the town square.[16] New England newspapers lampooned them in a poem written by Edward Coote Pinkney:

> *Why is that graceful female here*
> *With yon red hunter of the deer?*
> *Of gentle mien and shape, she seems*
> *For civil halls designed,*
> *Yet with the stately savage walks*
> *As she were of his kind.*

As soon as the Cherokee legislature declared marriages between white men and Cherokee women to be bigamy, in 1829, Elizabeth went to the Cherokee Supreme Court, sued James, and won. She was the first and only Cherokee woman to sue for bigamy in the short life of the Cherokee Supreme Court. She won the farm and five hundred dollars. It didn't do Elizabeth much good; nine years later, she had to leave the farm behind and go on the Trail of Tears and sit at Mantle Rock in the middle of winter.[17]

196   *Forgotten Landscapes*

WHAT WE CAN REGAIN

We can learn many things from Native American traditional life (chapter 9). But to fully embrace them, we need to also incorporate some Native American attitudes.

The Love of Finding Things Out

Natives entered the North American continent from Siberia and in less than a millennium had reached the southern tip of South America. They had to adapt their entire cultures, especially food and shelter, to an incredible range of environmental conditions. They could not depend on knowledge they brought with them from the steppes of Asia. They had to be closely observant of their environments and of all the plants and animals that shared it with them. This book is filled with numerous examples of the ingenious technology they developed.

In contrast, the white approach has often been not to adapt but to erase and replace. A major example of this is agricultural monocultures, as described in chapter 4.

Native cultures around the world are famous for being able to recognize hundreds of kinds of plants and animals, and for being able to read the landscape as they find their way across the countryside. Not so modern cultures, in which recreation is often indoors and food comes from stores. I have found from student notebooks, describing field trips, that they are often blind to nature. This is not limited to little animals that run and hide, but plant blindness as well.[18] They can look directly at blooming wildflowers and write, "There's nothing out there." Biodiversity (the diversity of species) is an important resource for us today. It was essential to the survival of prehistoric peoples. And they have to love knowing about wild plants and animals, an innate feeling that has been called *biophilia*.[19]

The white approach, in contrast, was that there was no need to know about the natural world. Of the countless examples I encounter almost daily, one sticks in my mind due to its sheer strangeness. As I explain in chapter 5, the present and future of the American West depends on water management. At the basis of water management, whether snowpack, river flow, or groundwater, one must know how much water there is.

In Oklahoma in 2024, two Republican legislators (in the party that frequently disdains environmental legislation) sponsored a bill that would require large farmers and industries to monitor how much groundwater they pump each year. The overwhelmingly Republican legislature passed

Epilogue 197

the bill, which places no limits on pumping and just requires monitoring. The governor, extreme even in his own Republican Party, vetoed it.[20] The message is very clear regarding water resources: We don't know, *and we don't want to know.* The Natives could not have settled America with that kind of attitude.

Reverence: Consider the Lilies

As she sat at Mantle Rock, Elizabeth knew that white Americans were accumulating treasures for themselves on Earth, and would stop at nothing to continue doing so. When she was in church—one of the Cherokee churches modeled on white Christian churches—Elizabeth heard that Jesus said to not accumulate treasures on Earth—"for where your treasure is, there will your heart be also." She even read these words in the Cherokee Bible translation published by a collaboration between white reverend Samuel Worcester and Cherokee scholar Elias Boudinot. And yet, she reflected, this is exactly what the white Americans were doing. Most Cherokees did not accumulate treasures on Earth. As a matter of fact, Cherokees did not even individually own land in the Cherokee Nation. It was the Americans who were living in a way that contradicted the teachings of the man they pretended was their savior.

Those learning a Native language frequently discover that for many tribes, there is no simple distinction between animate and inanimate objects. In English, something is either a person, an animal, a plant, or a thing. Most English speakers would classify plants as things. Some people consider animals to be things, while others consider them to be persons. But, as Louise Erdrich discussed in her novel *The Last Report on the Miracles at Little No Horse*, to a speaker of the Native American Anishinaabe language, stones and kettles can have spirits, and thus, to a certain extent, be animate. "I never thought of it that way before," say those trying to learn Navajo.[21]

Natives tend to look more closely at "things" and to notice more details about plants and animals. In this way, as in others, Natives tend to resemble the man Jesus, who said in the Sermon on the Mount (in the processed and reorganized version we have in the sixth chapter of the book of Matthew), "Consider the lilies of the field."

What does it mean to *consider*?

Most whites, including most of Jesus's fellow Jews thousands of years ago, have not ever *seen* the lilies of the field. ("Lilies," a translation of the Greek word *krinon*, may have referred to a number of different kinds of wildflower.) Oh, yeah, there are flowers out there, they might say after they

198   *Forgotten Landscapes*

are pointed out to them. Each individual lily was small, but collectively they covered the spring hillsides with a foam of color. Natives in the deserts of the Southwest and the foothills of California would have seen a set of plant species very similar to what Jesus saw.

Notice what Jesus did *not* say. He did not say to trample the lilies of the field, or to bulldoze them, or spray herbicide, pour oil, or dump trash on them. He did not say to glance at them and then go on your self-absorbed way. He did not say to argue about whether the plants evolved or were miraculously created, a topic on which I have wasted hundreds of hours and thousands of published words.

To consider the lilies, you have to walk lightly among them on your sandaled or moccasined feet. Then you have to crouch on the ground and look at them on their own level. In fact, look at just one of them, out of the tens of thousands on the hillside. Jesus did not say to pluck the lily so that you could stand up and examine at it on your own level, or add it to your pressed plant collection, the way I have had hundreds of students do.

I used to begin each semester of biology and botany classes in rural Oklahoma by writing a passage from one of Jesus's speeches on the board:

> *Even Solomon*
> *In all his glory*
> *Was not arrayed*
> *As one of these.*

I would then ask the students, most of whom had endured many hours of church and Sunday school, who said those words. Usually, nobody answered. A couple of times, someone would guess that Solomon said it. Just one time, a student replied that Jesus had said it. I followed up by asking what Jesus was talking about. This student said that he was talking about a wildflower. At least, that is the way I like to remember this one blessed time when a student got it right. That is, one wildflower, not the whole hillside.

The most glorious thing that Jesus or his listeners could imagine—the gleaming raiment of Israel's richest king—was less splendid than even just one wildflower out of thousands. This one lily displayed its beauty, close to the ground, not for us to see, but for its pollinators to see; and once the pollination season was over, the flower would wither away.

Jesus was very much like a Native observer—not that Jesus was the only one. Aristotle looked closely at organisms, too. This is not to say that all Natives, or even most, would notice a single wildflower. But it was

Epilogue   199

more consistent with the Native than with the white way of looking at the world, to notice the individualities and particularities of a single flower.

Western Christians often dismiss Native reverence for nature as idolatry. They like to imagine that Natives worship trees as gods. They even chuckle when they read about the prophet Isaiah performing a scientific experiment to disprove idolatry. Take a piece of wood from a tree, Isaiah said, and cut it in half. From half, make an idol. Burn the other half in the fireplace. There is no difference between the two halves. Ergo, the tree is not a god.

But I heard an Anishinaabe speaker at a conference in Minnesota who said his people revered trees as manifestations of the divine. "We don't think the tree is a god. We aren't stupid," he said. How many white Americans see a tree as a manifestation of the divine, and not just as a thing? Not enough.

Elizabeth knew that the solution to the problem was not a technological one. The Cherokees could not go back to the way life was before they adopted white customs. To solve the problems, the Americans would have to renounce their material greed, something that Elizabeth was certain could not happen.

By the twentieth century, many white Americans began to experience guilt for the conquest of the landscape and the Natives who lived there. When Elizabeth was an old woman, living in Fort Gibson, Indian Territory, she could see not only that the Cherokee portion of Indian Territory was far smaller than the Appalachian homeland, but also poor and dry. And she saw the land degrade during her lifetime. Did white civilization truly represent progress, or was it really the Darkening Land?

We could ask the same question today.

APPENDIX

Examples of white explorer accounts of Native Americans starting fires in North America, north of Mexico, either as observer or from informant, from Stewart[1] and from Bonnicksen,[2] except one reference from Sherow.[3] The habitat types are generalized, especially since forests, in this table, might refer to savannas, which would result from repeated fires. Does not include twentieth-century ethnographic studies. Total, 173.

Year	Location	Habitat	Observer/s
1524	North Carolina	Deciduous forest	Giovanni da Verrazzano
1542	Southwest North America	Savanna	Álvar Cabeza de Vaca
1555	Texas	Savanna	Álvar Cabeza de Vaca
1590	Virginia	Deciduous forest	John White
1602	Maine	Deciduous forest	Samuel Purchas
1605	Massachusetts	Deciduous forest	Samuel de Champlain
1607	Virginia	Deciduous forest	George Percy
1607	Virginia	Deciduous forest	Anonymous
1608	Virginia	Deciduous forest	John Smith
1609	Virginia	Deciduous forest	H. Spelman
1612	Virginia	Deciduous forest	William Strachey
1614	Virginia	Deciduous forest	Ralph Hamar
1624	Virginia	Deciduous forest	John Smith
1626	Massachusetts	Deciduous forest	Samuel Purchas

Year	Location	Habitat	Observer/s
1629	New England	Deciduous forest	William Wood
1630	Massachusetts	Deciduous forest	John Winthrop
1631	Delaware	Deciduous forest	Dutch explorer
1632	Massachusetts	Deciduous forest	Thomas Morton
1633	Delaware	Deciduous forest	David Pietersz. de Vries
1633	Potomac River	Deciduous forest	Andrew White
1634	Massachusetts	Deciduous forest	William Wood
1643	Rhode Island	Deciduous forest	Roger Williams
1649	Virginia	Deciduous forest	William Bullock
1650	Virginia	Deciduous forest	E. Williams
1654	Massachusetts	Deciduous forest	E. Johnson
1658	Quebec	Deciduous forest	Pierre Radisson
1670	Delaware	Deciduous forest	Daniel Denton
1670	Virginia	Deciduous forest	John Lederer
1679	Illinois	Savanna	Louis Hennepin
1683	Michigan	Deciduous forest	Louis Hennepin
1685	New York	Deciduous forest	Adriaen van der Donck
1685	Texas	Savanna	Henri Joutel
1691	Delaware	Deciduous forest	P. Lindstrom
1701	Virginia	Deciduous forest	F. L. Michel
1709	Virginia	Deciduous forest	John Lawson
1716	Virginia	Deciduous forest	Rev. James Fontaine
1718	Carolinas	Deciduous forest	John Lawson
1720	Ouachita River	Tallgrass prairie	LePage du Pratz
1722	Virginia	Deciduous forest	Robert Beverly
1728	Virginia, North Carolina	Deciduous forest	William Byrd
1731	Carolinas	Deciduous forest	Mark Catesby
1737	North Carolina	Deciduous forest	John Brickell
1748	Ohio	Deciduous forest	David McClure
1749	New Jersey	Deciduous forest	Peter Kalm

Year	Location	Habitat	Observer/s
1750	Ozark plateau	Savanna	Louis Vivier
1758	Mississippi River	Savanna	Le Page du Pratz
1760	New England	Deciduous forest	Andrew Burnaby
1763	Upper Mississippi River	Tallgrass prairie	Pierre-François de Charlevoix
1769	California	Oak savanna	Juan Crespí
1771	California	Coastal mix	Spanish mission report
1772	New York	Deciduous forest	William Campbell
1773	Florida	Savanna	William Bartram
1776	California	Oak savanna	Fernándo Rivera y Moncada
1776	Utah	Sagebrush, grassland	Domínguez–Escalante expedition
1776	California	Coastal mix	Pedro Font
1778	Upper Mississippi River	Tallgrass prairie	J. Carver
1788	Delaware	Deciduous forest	G. H. Loskiel
1790	Missouri	Deciduous forest	Captain Foucher
1792	Baja California	Desert	Jose Longinos Marinez
1792	California	Oak savanna	Jose Marinez
1794	Massachusetts	Deciduous forest	George Loskiel
1802	Lower Missouri River	Savanna	Lewis and Clark
1802	Kentucky	Deciduous forest	André Michaux
1803	New England	Deciduous forest	Isaac Weld
1804	Upper Missouri River	Shortgrass prairie	Lewis and Clark
1804	Upper Missouri River	Shortgrass prairie	Alexander Henry
1805	Wyoming	Shortgrass prairie	Lewis and Clark
1805	California	Oak woodland	H. Willis Baxley
1805	New York	Deciduous forest	T. Bigelow

204 *Forgotten Landscapes*

Year	Location	Habitat	Observer/s
1806	Montana, Wyoming	Coniferous forest	Lewis and Clark
1806	California	Oak savanna	Spanish expedition report
1806	Oregon	Coniferous forest	Lewis and Clark
1809	Nebraska	Tallgrass prairie	J. Bradbury
1812	Missouri	Deciduous forest	Amos Stoddard
1814	Missouri	Deciduous forest	H. M. Brackenridge
1818	Kentucky	Deciduous forest	James Flint
1818	California	Coastal mix	Otto von Kotzebue
1818	Illinois	Tallgrass prairie	E. P. Fordham
1819	Alleghenies	Deciduous forest	R. W. Wells
1819	Ohio	Deciduous forest	A. Bourne
1820	Missouri	Deciduous forest	H. R. Schoolcraft
1820	Minnesota	Deciduous forest	H. R. Schoolcraft
1820	Missouri	Deciduous forest	A. Bourne
1821	New England	Deciduous forest	Timothy Dwight
1821	Oklahoma	Deciduous forest	Thomas Nuttall
1822	New York	Deciduous forest	T. Dwight
1822	Texas	Savanna	Newspaper report
1823	New York	Deciduous forest	Duncan McMartin
1823	Oregon	Coniferous forest	David Douglas
1825	Illinois	Tallgrass prairie	Chester Loomis
1826	Rocky Mountains	Coniferous forest	Robert Campbell
1826	Missouri	Deciduous forest	T. Flint
1826	Oregon	Coniferous forest	David Douglas
1827	Oregon	Shortgrass prairie	Peter Ogden
1831	Idaho	Shortgrass prairie	Ross Cox
1832	Oklahoma	Savanna	Washington Irving
1832	Kansas	Tallgrass prairie	George Catlin
1832	Montana	Shortgrass prairie	Warren Ferris

Year	Location	Habitat	Observer/s
1832	Iowa, Nebraska	Tallgrass prairie	Prince Maximilian of Wied-Neuwied
1833	Idaho	Sagebrush, grassland	John K. Townsend
1834	Oregon	Shortgrass prairie	Nathaniel J. Wyeth
1835	Idaho	Shortgrass prairie	Osborne Russell
1835	Texas	Savanna	Herman Ehrenberg
1835	Wisconsin	Deciduous forest	C. F. Hoffman
1836	California	Coastal mix	Mariano Guadalupe Vallejo
1837	California	Coastal mix	HMS *Belcher* expedition
1837	Illinois	Tallgrass prairie	H. L. Ellsworth
1838	Illinois	Tallgrass prairie	A. D. Jones
1839	Idaho	Shortgrass prairie	John Work
1839	Idaho	Shortgrass prairie	T. J. Farnham
1840	Upper Missouri River	Shortgrass prairie	Pierre-Jean De Smet
1840	Idaho	Coniferous forest	Pierre-Jean De Smet
1841	Oregon	Coniferous forest	George Emmons
1841	Oregon	Coniferous forest	Joseph Drayton
1841	Oregon	Coniferous forest	Henry Eld
1843	Oregon	Coniferous forest	James Clyman
1844	Former Cherokee Nation	Deciduous forest	G. Featherstonhaugh
1844	San Joaquin Valley	Shortgrass prairie	P. B. Hinds
1844	Oklahoma	Savanna	Josiah Gregg
1846	Utah	Sagebrush, grassland	Edwin Bryant
1846	Utah	Sagebrush, grassland	Howard Egan
1848	San Joaquin Valley	Shortgrass prairie	Jacques A. Moerenhout
1848	San Joaquin Valley	Shortgrass prairie	E. Bryant

Year	Location	Habitat	Observer/s
1848	San Joaquin Valley	Shortgrass prairie	T. J. Farnham
1849	San Joaquin Valley	Shortgrass prairie	L. W. Hastings
1849	Alabama	Deciduous forest	Sir Charles Lyell
1850	California	Coniferous forest	Frank Marryat
1850	San Joaquin Valley	Shortgrass prairie	E. M. Kern
1850	San Joaquin Valley	Shortgrass prairie	L. A. Rice
1850	California	Coastal mix	William Redmond Ryan
1851	Iroquois land	Deciduous forest	L. H. Morgan
1851	Oregon	Coniferous forest	George Riddle
1851	California	Coniferous forest	George G. Gibbs
1851	California, Nevada	Sagebrush, grassland	Isabel Kelly
1854	California	Coastal mix	E. G. Beckwith
1854	Oklahoma	Savanna	A. W. Whipple
1854	Oregon	Shortgrass prairie	Pierre-Charles Saint-Amant
1855	California	Coastal mix	Frank Marryat
1855	California	Coniferous forest	Galen Clark
1857	Ohio	Deciduous forest	J. B. Finley
1857	Texas	Savanna	Frederick Law Olmsted
1857	Oregon	Prairie	M. Armstrong
1859	Kansas	Tallgrass prairie	S. N. Carvalho
1860	San Joaquin Valley	Shortgrass prairie	A. S. Taylor
1860	Idaho	Pine forest	P. M. Engle
1861	Yosemite	Coniferous forest	H. Willis Baxley
1863	Illinois	Tallgrass prairie	Henry Engleman
1867	Kentucky	Deciduous forest	John Muir
1868	Colorado	Coniferous forest	John Wesley Powell
1869	Illinois	Tallgrass prairie	J. D. Caton
1870	Kansas	Tallgrass prairie	George Sternberg
1873	Arizona	Coniferous forest	George Wheeler

Year	Location	Habitat	Observer/s
1873	Yosemite	Coniferous forest	Dennis Kane
1874	Minnesota	Deciduous forest	J. A. Allen
1875	Oregon	Coastal mix	Charles Nordhoff
1875	Washington, Montana	Shortgrass prairie	H. H. Bancroft
1875	San Joaquin Valley	Shortgrass prairie	H. H. Bancroft
1877	Kansas	Shortgrass prairie	R. I. Dodge
1878	Massachusetts	Deciduous forest	Joseph H. Brown
1879	Missouri	Deciduous forest	G. C. Swallow
1879	Colorado	Coniferous forest	W. L. Campbell
1879	Canada	Coniferous forest	George M. Davidson
1880	Oregon	Coniferous forest	Samuel Clarke
1882	Yosemite	Coniferous forest	M. C. Briggs
1883	California	Coniferous forest	Gordon Cumming
1884	Kentucky	Deciduous forest	John Hussey
1884	Minnesota	Deciduous forest	W. J. McGee
1885	Alaska	Boreal forest	Henry Allan
1887	California	Coniferous forest	Joaquin Miller
1891	Michigan	Deciduous forest	N. S. Shaler
1894	Yosemite	Coniferous forest	Galen Clark
1894	Sierra Nevada	Coniferous forest	John Muir
1894	Virginia	Deciduous forest	Gerard Fowke
1897	Illinois, Indiana, Iowa	Tallgrass prairie	C. S. Sargent

NOTES

INTRODUCTION

1. Blackburn, Bob, Duane King, and Neil Morton. *Cherokee Nation: A History of Survival, Self Determination, and Identity*. Tahlequah, OK: Cherokee Nation, 2018. Ehle, John. *Trail of Tears: The Rise and Fall of the Cherokee Nation*. New York: Doubleday, 1988. Fitzgerald, David C., and Duane King. *The Cherokee Trail of Tears*. Portland, OR: Graphic Arts Books, 2007. Thornton, Russell. *American Indian Holocaust and Survival: A Population History since 1492*. Norman: University of Oklahoma Press, 1990. Weatherford, Jack. *Native Roots: How the Indians Enriched America*. New York: Crown, 1991. Woodward, Grace Steele. *The Cherokees*. Norman: University of Oklahoma Press, 1959.

2. Perdue, Theda. *Cherokee Women: Gender and Culture Change, 1700–1835*. Lincoln: University of Nebraska Press, 1998. This work specifically identifies the plight of Elizabeth Hilderbrand Pettit.

3. Crosby, Alfred W. *Ecological Imperialism: The Biological Expansion of Europe, 900–1900*. New York: Cambridge University Press, 1986.

CHAPTER 1

1. More, Sir Thomas. *Utopia*. Translated by Dominic Baker-Smith. New York: Penguin, 2012.

2. Bregman, Rutger. *Utopia for Realists: How We Can Build the Ideal World*. Translated by Elizabeth Manton. New York: Little, Brown, 2014, 2017.

3. Heyer, Évelyne. *L'Odyssée des Gènes*. Paris, France: Flammarion, 2020.

4. Koehler, Heidi, and Alexandra Weil. "Study confirms age of oldest fossil human footprints in North America." United States Geological Survey, October 5,

210   *Forgotten Landscapes*

2023. Available online. URL: https://www.usgs.gov/news/national-news-release /study-confirms-age-oldest-fossil-human-footprints-north-america. Accessed October 24, 2023.

5. Alex, Bridget. "Monte Verde: Our earliest evidence of humans living in South America." *Discover*, November 1, 2019. Available online. URL: https:// www.discovermagazine.com/planet-earth/monte-verde-our-earliest-evidence-of -humans-living-in-south-america. Accessed September 12, 2023.

6. Erlandson, Jon M., Torben C. Rick, Todd J. Braje, Molly Casperson, Brendan Culleton, Brian Fulfrost, Tracy Garcia, Daniel A. Guthrie, Nicholas Jew, and Lauren Willis. "Paleoindian seafaring, maritime technologies, and coastal foraging on California's Channel Islands." *Science* 331 (2011): 1,181–1,185.

7. Stuart, Anthony J. *Vanished Giants: The Lost World of the Ice Age.* Chicago: University of Chicago Press, 2022.

8. Mason, Gregory. *Columbus Came Late.* New York: Century, 1931.

9. Diamond, Jared. *Collapse: How Societies Choose to Fail or Succeed.* New York: Viking, 2005.

10. Nash, Lenna M., and Eve A. Hargrave. "Human sacrifice in the Late Prehistoric American Bottom: Skeletal and archaeological evidence." University of Illinois, 2015. Available online. URL: https://experts.illinois.edu/en/publications /human-sacrifice-in-the-late-prehistoric-american-bottom-skeletal-. Accessed August 10, 2023.

11. BBC.com. "Spiro Mounds: North America's Lost Civilisation." Available online. URL: https://www.bbc.com/travel/article/20210621-spiro-mounds -north-americas-lost-civilisation. Accessed June 8, 2023.

12. Bartram, William. *Travels through North and South Carolina, Georgia, East and West Florida.* Savannah, GA: Facsimile of the 1792 London edition published 1973 by Beehive Press.

13. Cronley, Connie. *A Life on Fire: Oklahoma's Kate Barnard.* Norman: University of Oklahoma Press, 2021, p. 159.

14. Sherow, James E. *The Grasslands of the United States: An Environmental History.* Santa Barbara, CA: ABC-CLIO, 2007.

15. Burford, E. J. *London: The Synfulle Citie.* London, UK: Robert Hale, 1990.

16. Crosby, Alfred W. *Ecological Imperialism: The Biological Expansion of Europe, 900–1900.* New York: Cambridge University Press, 1986. Fenn, Elizabeth A. *Pox Americana: The Great Smallpox Epidemic of 1775–82.* New York: Farrar, Straus and Giroux, 2001.

17. Matossian, Mary Kilbourne. *Poisons of the Past: Molds, Epidemics, and History.* New Haven, CT: Yale University Press, 1991.

18. Height is just one indicator of physical health, but is one that can be measured from archaeological specimens. Men of the Plains tribes averaged five feet, eight inches in height, compared to five feet, five inches for Europeans in the seventeenth and eighteenth centuries. Prince, Joseph M., and Richard H. Steckel.

"Tallest in the world: Native Americans of the Great Plains in the nineteenth century." National Bureau of Economic Research (December 1998). Vikings were taller, but not necessarily healthier, than other Europeans. European Americans, better fed than Europeans, were about the same height as Native Americans.

19. Demos, John. *The Unredeemed Captive: A Family Story from Early America.* New York: Knopf, 1994.

20. Balter, Michael. "Ancient DNA links Native Americans with Europe." *Science* 342 (October 25, 2013): 409–410.

21. Wade, Lizzie. "Ancient DNA confirms Native Americans' deep roots in North and South America." *Science* (November 8, 2018).

22. Pompeani, David P., Broxton W. Bird, Jeremy J. Wilson, William P. Gilhooly III, Aubrey L. Hillman, Matthew S. Finkenbinder, and Mark B. Abbott. "Severe Little Ice Age drought in the midcontinental United States during the Mississippian abandonment of Cahokia." *Scientific Reports* 11 (2021). Available online. URL: https://www.nature.com/articles/s41598-021-92900-x. Accessed July 13, 2023. White, A. J., Samuel E. Munoz, Sissel Schroeder, and Lora R. Stevens. "After Cahokia: Indigenous repopulation and depopulation of the Horseshoe Lake Watershed AD 1400–1900." *American Antiquity* 85 (2020): 263–278. Available online. URL: https://www.cambridge.org/core/journals/american-antiquity/article/after-cahokia-indigenous-repopulation-and-depopulation-of-the-horseshoe-lake-watershed-ad-14001900/C678D62BE9FAB07C9283D62BE9757685. Accessed August 10, 2023.

23. King, Duane H. *Emissaries of Peace: The 1762 Cherokee and British Delegations.* Cherokee, NC: Museum of the Cherokee Indian Press, 2006. King, Duane H., ed. *The Memoirs of Lt. Henry Timberlake: The Story of a Soldier, Adventurer, and Emissary to the Cherokees, 1756–1765.* Cherokee, NC: Museum of the Cherokee Indian Press, 2007. Rogers, Anne F., and Barbara R. Duncan, eds. *Culture, Crisis & Conflict: Cherokee British Relations 1756–1765.* Cherokee, NC: Museum of the Cherokee Indian Press, 2009.

24. Little, Becky. "The Native American government that helped inspire the United States Constitution." *The History Channel,* 2020. Available online. URL: https://www.history.com/news/iroquois-confederacy-influence-us-constitution. Accessed July 6, 2023.

25. Hammerstedt, Scott W., and Amanda L. Regnier. *Spiro Mounds and WPA Archaeology in Oklahoma.* Charleston, SC: Arcadia Publishing, 2023.

26. Thulin, Lila. "Racist phrase found etched on Native American petroglyphs in Utah." *Smithsonian* (April 30, 2021). Available online. URL: https://www.smithsonianmag.com/smart-news/vandals-deface-native-american-petroglyphs-utah-racist-phrase-180977630/. Accessed December 30, 2023.

27. Cowley, Geoffrey, Andrew Mure, Nonny de la Pena, and Vicki Quade. "The plunder of the past." *Newsweek* (June 26, 1989), 58–60.

212   *Forgotten Landscapes*

CHAPTER 2

1. Suttie, J. M., S. G. Reynolds, and C. Batello. *Grasslands of the World*. Rome: Food and Agriculture Organization of the United Nations, 2005. Available online. URL: https://www.fao.org/3/y8344e/y8344e00.htm#Contents. Accessed July 3, 2023.

2. Rice, Stanley A., and Sonya L. Ross. "Smoke-induced germination in *Phacelia strictiflora*." *Oklahoma Native Plant Record* 13 (2013).

3. Egan, Timothy. *The Big Burn: Teddy Roosevelt and the Fire That Saved America*. Boston: Houghton Mifflin, 2009.

4. Francaviglia, Richard V. *The Cast Iron Forest: A Natural and Cultural History of the North American Cross Timbers*. Austin: University of Texas Press, 2000.

5. Flores, D. "The Great Plains 'wilderness' as a human-shaped environment." *Great Plains Research* 9 (1999): 343–355.

6. Delcourt, Hazel R., and Paul A. Delcourt. "Pre-Columbian Native American use of fire on southern Appalachian landscapes." *Conservation Biology* 11 (1997): 1,010–1,014.

7. Stewart, Omer Call. *Forgotten Fires: Native Americans and the Transient Wilderness*. Original title: *The Effects of Burning of Grasslands and Forests by Aborigines the World Over*. Edited by Henry T. Lewis and M. Kat Anderson. Norman: University of Oklahoma Press, 2002.

8. Bonnicksen, Thomas M. *America's Ancient Forests: From the Ice Age to the Age of Discovery*. New York: John Wiley, 2000. See also Pyne, S. J. *Fire in America: A Cultural History of Wildland and Rural Fire*. Princeton, NJ: Princeton University Press, 1982.

9. Anderson, M. Kat. "An Ecological Critique." In: Stewart, Omer Call, *Forgotten Fires: Native Americans and the Transient Wilderness*. Norman: University of Oklahoma Press, 2002.

10. Williams, Alec. "'We're falling into a ring of fire: Taking stock of wildfire liability regimes from varying perspectives in the United States." *Georgetown Environmental Law Review* (March 13, 2021). Available online. URL: https://www.law.georgetown.edu/environmental-law-review/blog/were-falling-into-a-ring-of-fire-taking-stock-of-wildfire-liability-regimes-from-varying-perspectives-in-the-united-states/. Accessed July 2, 2023.

11. Sherow, James E. *Grasslands of the United States: An Environmental History*. Santa Barbara, CA: ABC-CLIO, 2007.

12. Bartram, William. *Travels through North and South Carolina, Georgia, East and West Florida*. Savannah, GA: Beehive Press, facsimile of 1792 London edition.

13. Carter, Vachel A., Andrea Brunelle, Mitchell J. Power, R. Justin DeRose, Matthew F. Bekker, Isaac Hart, Simon Brewer, Jerry Spangler, Erick Robinson, Mark Abbott, S. Yoshi Maezumi, and Brian F. Codding. "Legacies of Indigenous land use shaped past wildfire regimes in the basin-plateau region, USA." *Communi-*

cations *Earth & Environment* 2 (2021). Available online. URL: https://www.nature.com/articles/s43247-021-00137-3. Accessed July 3, 2023.

14. Stephens, Scott L., and William J. Libby. "Anthropogenic fire and bark thickness in coastal and island pine populations from Alta and Baja California." *Journal of Biogeography* 33 (2006): 648–652.

15. Keeley, Jon E. "Native American impacts on fire regimes of the California coastal ranges." *Journal of Biogeography* 29 (2002): 303–320.

16. Anderson, M. Kat. "The fire, pruning, and coppice management of temperate ecosystems for basketry material by California Indian tribes." *Human Ecology* 27 (1999): 79–113. Anderson, M. Kat. "Uses of fire by Native Americans in California." Chapter 17 in Sugihara, Neil G., et al., eds., *Fire in California's Ecosystems.* Berkeley: University of California Press, 2006.

17. Roos, Christopher I., Thomas W. Swetnam, T. J. Ferguson, and Christopher A. Kiahtipes. "Native American fire management at an ancient wildland–urban interface in the southwest United States." *Proceedings of the National Academy of Sciences USA,* January 18, 2021. Available online. URL: https://www.pnas.org/doi/abs/10.1073/pnas.2018733118. Accessed July 3, 2023.

18. King, James E. "Late Quaternary vegetational history of Illinois." *Ecological Monographs* 51 (1981): 43–62. An image of an Illinois State Museum display based on King's data is available at https://exhibits.museum.state.il.us/exhibits/midewin/pollendiag.html. Accessed August 19, 2023.

19. Rice, Stanley A. *Scientifically Thinking: How to Liberate Your Mind, Solve the World's Problems, and Embrace the Beauty of Science.* Amherst, NY: Prometheus Books, 2018.

20. Chavennes, Elizabeth. "Written records of forest succession." *Scientific Monthly* 53 (1941): 76–77. Gleason, Henry Allen. "The relation of forest distribution and prairie fires in the middle west." *Torreya* 13 (1913): 173–181.

21. Ehle, John. *Trail of Tears: The Rise and Fall of the Cherokee Nation.* New York: Anchor, 1988. Rozema, Vicki. *Voices from the Trail of Tears.* Winston-Salem, NC: John F. Blair, 2003.

22. Lewis, Henry T. "An Anthropological Critique." In: Stewart, Omer Call, *Forgotten Fires: Native Americans and the Transient Wilderness.* Norman: University of Oklahoma Press, 2002.

23. Rice, Stanley A. "Tundra." In *Encyclopedia of Biodiversity.* New York: Facts on File, 2012.

24. Barbour, Michael, Bruce Pavlik, Frank Drysdale, and Susan Lindstrom. *California's Changing Landscapes: Diversity and Conservation of California's Vegetation.* Sacramento: California Native Plant Society, 1993.

25. Shive, Kristen, Christy Brigham, Tony Caprio, and Paul Hardwick. "2021 fire season impacts to giant sequoias." National Park Service. Available online. URL: https://www.nps.gov/articles/000/2021-fire-season-impacts-to-giant-sequoias.htm. Accessed December 30, 2023.

214   *Forgotten Landscapes*

26. Koch, Alexander, Chris Brierley, Mark M. Maslin, and Simon L. Lewis. "Earth system impacts of the European arrival and Great Dying in the Americas after 1492." *Quaternary Science Reviews* 207 (2019): 13–36.

27. Borden, Caroline G., Marlyse C. Duguid, and Mark S. Ashton. "The legacy of fire: Long-term changes to the forest understory from periodic burns in a New England oak-hickory forest." *Fire Ecology* 17 (2021). Available online. URL: https://fireecology.springeropen.com/articles/10.1186/s42408-021-00115-2. Accessed July 3, 2023. Klimaszewski-Patterson, Anna, and Scott Mensing. "Paleoecological and paleolandscape modeling support for pre-Columbian burning by Native Americans in the Golden Trout Wilderness Area, California, USA." *Landscape Ecology* 35 (2020): 2,659–2,678.

CHAPTER 3

1. Schorger, A. W. *The Passenger Pigeon: Its Natural History and Extinction.* Madison: University of Wisconsin Press, 1955.

2. Yale University, "The lost birds of Audubon." Available online. URL: https://beineckeaudubon.yale.edu/news/lost-birds-audubon. Accessed April 6, 2024.

3. Department of Vertebrate Zoology, National Museum of Natural History. "Spotlight: The passenger pigeon." Available online. URL: https://www.si.edu/spotlight/passenger-pigeon. Accessed April 4, 2024.

4. Novak, B. J., J. A. Estes, H. E. Shaw, E. V. Novak, and B. Shapiro. "Experimental investigation of the dietary ecology of the extinct passenger pigeon, *Ectopistes migratorius.*" *Frontiers in Ecology and Evolution* 6 (2018). Available online. URL: https://www.frontiersin.org/articles/10.3389/fevo.2018.00020/full. Accessed January 21, 2025. Ellsworth, J. W., and B. C. McComb. "Potential effects of passenger pigeon flocks on the structure and composition of presettlement forests of eastern North America." *Conservation Biology* 17 (2003): 1,548–1,558.

5. Guiry, Eric J., Trevor J. Orchard, Thomas C. A. Royle, Christina Cheung, and Dongya Y. Yang. "Dietary plasticity and the extinction of the passenger pigeon (*Ectopistes migratorius*)." *Quaternary Science Reviews* 233 (2020): 106225. Available online. URL: https://www.sciencedirect.com/science/article/abs/pii/S0277379119306857. Accessed April 4, 2024.

6. Hung, Chih-Ming, Pei-Jen Shaner, Robert M. Zink, Wei-Chung Liu, Te-chin Chu, Wen-San Huang, and Shou-shien Li. "Drastic population fluctuations explain the rapid extinction of the passenger pigeon." *Proceedings of the National Academy of Sciences* 111 (2014): 10,636–10,641.

7. Neumann, Thomas W. "Human–wildlife competition and the passenger pigeon: Population growth from system destabilization." *Human Ecology* 13 (1985): 389–410.

8. Orchard, Trevor J., Suzanne Needs-Howarth, Alicia L. Hawkins, Louis Lesage, Eric J. Guiry, and Thomas C. A. Royle. "Considering passenger pigeon abun-

Notes 215

dance and distribution in the Late Woodland zooarchaeological record of southern Ontario, Canada." *International Journal of Osteoarchaeology* 33 (2023): 608–618.

9. DeCalesta, David S. *Predator Control: History and Policies.* Oregon State University Extension Service Circular 710. Available online. URL: https://digitalcollections.library.oregon.gov/nodes/view/285950. Accessed January 21, 2025.

10. One of many places that this quote has been posted online is the Sierra Club. "Thinking Like a Mountain." Available online. URL: https://www.sierraclub.org/sites/www.sierraclub.org/files/sce/rocky-mountain-chapter/Wolves-Resources/Thinking%20Like%20a%20Mountain%20-%20Aldo%20Leopold.pdf. Accessed April 6, 2024.

11. Binkley, Dan, Margaret M. Moore, William H. Romme, and Peter Brown. "Was Aldo Leopold right about the Kaibab deer herd?" *Ecosystems* 9 (2006): 227–241.

12. Bartram, William. *Travels through North and South Carolina, Georgia, East and West Florida. Savannah, Georgia.* Bronx, NY: Beehive Press, 1973. Facsimile of the 1792 edition.

13. Schwartz, Joe. "The ten plagues: Natural disasters or divine intervention?" McGill University Office for Science and Society, 2019. Available online. URL: https://www.mcgill.ca/oss/article/ten-plagues-environmental-disasters-or-religious-interference. Accessed March 30, 2024.

14. Gross, Michael. "How locusts become a plague." *Current Biology* 31 (2021): R459–R461. Available online. URL: https://www.sciencedirect.com/science/article/pii/S0960982221006667. Accessed March 30, 2024.

15. RoadsideAmerica.com. "The Great Mouse Invasion—West Kern Oil Museum." Available online. URL: https://www.roadsideamerica.com/story/36898. Accessed August 21, 2023. Pearson, Oliver P. "History of two local outbreaks of feral house mice." *Ecology* 44 (1963): 540–549.

16. Kim, Seong-Hee, John Tschirhart, and Steven W. Buskirk. "Reconstructing past population processes with general equilibrium models: House mice in Kern County, California, 1926–1927." *Ecological Modelling* 209 (2007): 235–248.

17. Stewart, Omer C. *Forgotten Fires.* Edited by Henry T. Lewis and M. Kat Anderson. Norman: University of Oklahoma Press, 2002, p. 74.

18. Bonnicksen, Thomas M. *America's Ancient Forests: From the Ice Age to the Age of Discovery.* New York: Wiley, 2000.

19. Bonnicksen, *op. cit.*

20. Schorger, *op. cit.*

21. Philip, John. *A Law of Blood: The Primitive Law of the Cherokee Nation.* DeKalb: Northern Illinois University Press, 1970.

22. Bechmann, Roland. *Trees and Man: The Forest in the Middle Ages.* New York: Paragon House, 1990.

23. Sierra Club, *op. cit.*

24. I include a list of nineteen examples in Rice, Stanley A., *Encyclopedia of Evolution.* New York: Facts on File, 2006, 317.

216 *Forgotten Landscapes*

25. Barnosky, Anthony D. "Assessing the causes of late Pleistocene extinctions on the continents." *Science* 306 (2004): 70–75.

26. Martin, Paul S., and R. G. Klein. *Quaternary Extinctions: A Prehistoric Revolution.* Tempe: University of Arizona Press, 1984.

27. O'Keefe, F. Robin, et al. "Pre–Younger Dryas megafaunal extirpation at Rancho La Brea linked to fire-driven state shift." *Science* 381, no. 6699 (2023). Available online. URL: https://www.science.org/doi/10.1126/science.abo3594. Accessed April 5, 2024.

28. Firestone, R. B., et al. "Evidence for an extraterrestrial impact 12,900 years ago that contributed to the megafaunal extinctions and the Younger Dryas cooling." *Proceedings of the National Academy of Sciences* 104 (2007): 16,016–16,021.

CHAPTER 4

1. Blake, Michael. *Maize for the Gods: Unearthing the 9,000-Year History of Corn.* Berkeley: University of California Press, 2015. Mann, Charles C. "What ancient maize can tell us about thousands of years of civilization in America." *Smithsonian Magazine* (November 2018). Available online. URL: https://www.smithsonianmag.com/smithsonian-institution/ancient-maize-thousands-years-civilization-america-180970543/. Accessed July 6, 2023. Weatherford, Jack. *Indian Givers: How the Indians of America Transformed the World.* New York: Fawcett Columbine, 1988.

2. Sherow, James E. *Grasslands of the United States: An Environmental History.* Santa Barbara, CA: ABC-CLIO, 2007.

3. Thornton, Russell. *The Cherokees: A Population History.* Lincoln: University of Nebraska Press, 1990. Ehle, John. *Trail of Tears: The Rise and Fall of the Cherokee Nation.* New York: Anchor, 1988.

4. Bess, Michael. *The Light-Green Society: Ecology and Technological Modernity in France, 1960–2000.* Chicago: University of Chicago, 2003.

5. Cook, Christopher D. *Diet for a Dead Planet: How the Food Industry Is Killing Us.* New York: The New Press, 2004.

6. Malmström, Helena, Anna Linderholm, Pontus Skoglund, Jan Storå, Per Sjödin, M. Thomas P. Gilbert, Gunilla Holmlund, Eske Willerslev, Mattias Jakobsson, Kerstin Lidén, and Anders Götherström. "Ancient mitochondrial DNA from the northern fringe of the Neolithic farming expansion in Europe sheds light on the dispersion process." *Philosophical Transactions of the Royal Society B* 370 (2015). Available online. URL: https://royalsocietypublishing.org/doi/full/10.1098/rstb.2013.0373. Accessed June 29, 2023.

7. Lopez-Ridaura, Santiago, Luis Barba-Escoto, Cristian A. Reyna-Ramirez, Carlos Sum, Natalia Palacios-Rojas, and Bruno Gerard. "Maize intercropping in the milpa system. Diversity, extent and importance for nutritional security in the Western Highlands of Guatemala." *Scientific Reports* 11 (2021). Available online. URL: https://www.nature.com/articles/s41598-021-82784-2, Accessed June 14, 2023.

Notes   217

8. Jackson, Dana L., and Laura L. Jackson, eds. *The Farm as Natural Habitat: Reconnecting Food Systems with Ecosystems.* Covelo, CA: Island Press, 2002.

9. Kimmerer, Robin Wall. *Braiding Sweetgrass: Indigenous Wisdom, Scientific Knowledge, and the Teachings of Plants.* Vancouver, British Columbia: Milkweed Editions, 2013. Ngapo, T. M., P. Bilodeau, Y. Arcand, et al. "Historical Indigenous food preparation using produce of the Three Sisters intercropping system." *Foods* 10, no. 524 (2021). Available online. URL: https://www.mdpi.com/2304-8158/10/3/524. Accessed June 14, 2023.

10. Pat Gwin, personal communication, 2023.

11. Mann, Charles C. "We are here. North America's Indigenous nations are reclaiming their sovereignty: Control of their land, laws, and how they live." *National Geographic* (July 2022): 36–75.

12. Kimmerer, *op. cit.*

13. Maathai, Wangari. *Unbowed: A Memoir.* New York: Random House, 2006.

14. Rice, Stanley A. *Green Planet: How Plants Keep the Earth Alive.* New Brunswick, NJ: Rutgers University Press, 2009.

15. Cocannouer, Joseph A. *Weeds, Guardians of the Soil.* Old Greenwich, CT: Devin-Adair, 1950.

CHAPTER 5

1. Zocchi, Druve, Gunther Wennemuth, and Yuki Ohta. "The cellular mechanism for water detection in the mammalian taste system." *Nature Neuroscience* 20 (2017): 927–933.

2. Doebley, John F. "'Seeds' of wild grasses: A major food of the southwestern Indians." *Economic Botany* 38 (1984): 52–64.

3. Hundley, Norris, Jr. *The Great Thirst: Californians and Water, 1770s–1990s.* Berkeley: University of California Press, 1992, 17.

4. Steward, Julian H. "Irrigation without agriculture." *Papers of the Michigan Academy of Sciences, Arts and Letters* 12 (1930): 149–156.

5. Lekson, Stephen H., Thomas C. Windes, John R. Stein, and W. James Judge. "The Chaco Canyon Community." *Scientific American* 259 (1988): 100–109.

6. Short, Henry L. "The deterioration of the Pueblo Bonito Great House in the Chaco Culture National Historical Park, New Mexico, USA." *PLOS One* 17 (2022): e0266099. Available online. URL: https://journals.plos.org/plosone/article?id=10.1371/journal.pone.0266099. Accessed April 27, 2024.

7. Guiterman, Christopher H., Thomas W. Swetnam, and Jeffrey S. Dean. "Eleventh-century shift in timber procurement areas for the great houses of Chaco Canyon." *Proceedings of the National Academy of Sciences* 113 (2015): 1,186–1,190. Nash, Stephen E. "Where did Chaco Canyon's timber come from?" *Smithsonian* (April 17, 2023). Available online. URL: https://www.smithsonianmag.com/science-nature/where-did-chaco-canyons-timber-come-from-180981996/. Accessed May 2, 2024.

218 *Forgotten Landscapes*

8. Betancourt, Julio L., Jeffrey S. Dean, and Herbert M. Hull. "Prehistoric long-distance transport of construction beams, Chaco Canyon, New Mexico." *American Antiquity* 51 (1986): 370–375.

9. Reynolds, Amanda C., Julio L. Betancourt, Jay Quade, P. Jonathan Patchett, Jeffrey S. Dean, and John Stein. "87Sr/86Sr sourcing of ponderosa pine used in Anasazi great house construction at Chaco Canyon, New Mexico." *Journal of Archaeological Science* 32 (2005): 1,061–1,075.

10. Benson, Larry V., and Deanna M. Grimstead. "Prehistoric Chaco Canyon, New Mexico: Residential population implications of limited agricultural and mammal productivity." *Journal of Archaeological Science* 108 (2019): 104971. Available online. URL: https://www.sciencedirect.com/science/article/abs/pii /S0305440319300603. Accessed April 7, 2024.

11. Scarborough, Vernon L., Samantha G. Fladd, Nicholas P. Dunning, et al. "Water uncertainty, ritual predictability and agricultural canals at Chaco Canyon, New Mexico." *Antiquity* 92 (2018): 870–889.

12. Wills, W. H., Brandon L. Drake, and Weatherbee B. Dorshow. "Prehistoric deforestation at Chaco Canyon?" *Proceedings of the National Academy of Sciences* 111 (2014): 11,584–11,591. Lentz, David L., et al. "Ecosystem impacts by the ancestral Puebloans of Chaco Canyon, New Mexico, USA." *PLOS One* (2021). Available online. URL: https://doi.org/10.1371/journal.pone.0258369. Accessed April 7, 2024. See also chapter 4 of Diamond, Jared. *Collapse: How Societies Choose to Fail or Succeed.* New York: Viking, 2005.

13. Betancourt, Julio L., Thomas R. Van Devender, and Paul S. Martin. *Packrat Middens: The Last 40,000 Years of Biotic Change.* Tempe: University of Arizona Press, 1990.

14. Roberts, David. "Riddles of the Anasazi." *Smithsonian* (July 2003). Available online. URL: https://www.smithsonianmag.com/history/riddles-of-the -anasazi-85274508/. Accessed April 29, 2024. Turner, Christy G. II, and Jacqueline A. Turner. *Man Corn: Cannibalism and Violence in the Prehistoric American Southwest.* Salt Lake City: University of Utah Press, 1998. Marlar, Richard A., Banks L. Leonard, Brian R. Billman, Patricia M. Lambert, and Jennifer E. Marlar. "Biochemical evidence of cannibalism at a prehistoric Puebloan site in southwestern Colorado." *Nature* 407 (2000): 74–78.

15. Sherow, James E. *Grasslands of the United States: An Environmental History.* Santa Barbara, CA: ABC-CLIO, 2007.

16. DeMola, Paul G. "The Hohokam: Canal masters of the American Southwest." *Popular Archaeology* (Winter 2019). Available online. URL: https:// popular-archaeology.com/article/the-hohokam-canal-masters-of-the-american -southwest/. Accessed April 29, 2024.

17. Hunt, Robert C., David Guillet, David R. Abbott, James Bayman, Paul Fish, Suzanne Fish, Keith Kintigh, and James A. Neely. "Plausible ethnographic analogies for the social organization of Hohokam canal irrigation." *American Antiq-*

uity 70 (2005): 433–456. Rice, Glen. "War and water: An ecological perspective on Hohokam irrigation." *Kiva* 63 (1998): 263–301.

18. Fish, Suzanne K., and Paul R. Fish. "Prehistoric desert farmers of the Southwest." *Annual Review of Anthropology* 23 (1994): 83–108.

19. Woodson, M. Kyle, Jonathan A. Sandor, Colleen Strawhacker, and Wesley D. Miles. "Hohokam canal irrigation and the formation irragric anthrosols in the Middle Gila River Valley, Arizona." *Geoarchaeology* 30 (2015): 271–290.

20. Rice, Stanley A. "Agriculture, evolution of." *Encyclopedia of Evolution.* New York: Facts on File, 2006.

21. Al-Ansari, Nadhir, Nasrat Adamo, and Varoujan K. Sissakian. "Hydrological characteristics of the Tigris and Euphrates Rivers." *Journal of Earth Sciences and Geotechnical Engineering* 9 (2019): 1–26.

22. Jacobsen, Thorkild, and Robert M. Adams. "Salt and silt in ancient Mesopotamian agriculture: Progressive changes in soil salinity and sedimentation contributed to the breakup of past civilizations." *Science* 128 (1958): 1,251–1,258.

23. Hillel, Daniel. *Out of the Earth: Civilization and the Life of the Soil.* Berkeley: University of California, 1992.

24. Ortloff, Charles R. "The water supply and distribution system of the Nabataean city of Petra (Jordan), 300 BC–300 AD." *Cambridge Archaeological Journal* 15 (2005): 93–109.

25. Fishman, Charles. *The Big Thirst: The Secret Life and Turbulent Future of Water.* New York: Free Press, 2011.

26. Trotta, Daniel, and Brad Brooks. "Western states reach 'historic' deal to help save Colorado River." Reuters, May 23, 2023. Available online. URL: https://www.reuters.com/world/us/us-states-reach-colorado-river-water-conservation-deal-interior-dept-2023-05-22/. Accessed May 1, 2024.

27. Hundley, Norris, Jr. *The Great Thirst: Californians and Water, 1770s–1990s.* Berkeley: University of California, 1992.

28. Reisner, Marc. *Cadillac Desert: The American West and Its Disappearing Water.* New York: Viking, 1986.

29. Fowler, Catherine S., and Nancy Peterson Walter. "Harvesting pandora moth larvae with the Owens Valley Paiute." *Journal of California and Great Basin Anthropology* 7 (1985): 155–165. Mitton, Jeff. "Brine flies provide a feast for birds and humans." *Colorado Arts and Sciences Magazine* (January 2012). Available online. URL: https://www.colorado.edu/asmagazine-archive/node/892. Accessed May 1, 2024.

CHAPTER 6

1. Abrams, Marc D., and Gregory J. Nowacki. "Native Americans as active and passive promoters of mast and fruit trees in the eastern USA." *The Holocene*

Forgotten Landscapes

18 (2008): 1,123–1,137. Available online. URL: https://journals.sagepub.com/doi/10.1177/0959683608095581. Accessed June 25, 2023.

2. Miami University of Ohio. Empire and American Religion. "Eulogy on King Philip, William Apess, 1836." Available online. URL: https://sites.miamioh.edu/empire/files/2022/08/1836-Apess-Eulogy-on-King-Philip.pdf. Accessed August 29, 2023.

3. Berenbaum, May. "Reality bites." *American Entomologist* 64: 134–137 (2018). Available online. URL: https://academic.oup.com/ae/article/64/3/134/5098349?login=false. Accessed June 25, 2023.

4. McAfee, Alison. "The problem with honey bees." *Scientific American* (November 2020). Available online. URL: https://www.scientificamerican.com/article/the-problem-with-honey-bees/. Accessed June 25, 2023.

5. Oyenini, Doyen. "Forget Cattle Rustling—Watch Out for Bee Rustlers." *Texas Monthly* (January 11, 2017). Available online. URL: https://www.texasmonthly.com/the-daily-post/bee-rustlers-strike-beekeeper-manvel/. Accessed June 25, 2023.

6. Brown, Dee. "The Long Walk of the Navahos." Chapter 2 of *Bury My Heart at Wounded Knee: An Indian History of the American West*. New York: Bantam Books, 1971.

7. Bonnicksen, Thomas M. *America's Ancient Forests: From the Ice Age to the Age of Discovery.* New York: John Wiley, 2000.

8. Keener, Craig, and Erica Kuhns. "The impact of Iroquoian populations on the northern distribution of pawpaws in the Northeast." *North American Archaeologist* 18 (1998): 327–342.

9. Cypher, Brian L., and Ellen A. Cypher. "Germination rates of tree seeds ingested by coyotes and raccoons." *American Midland Naturalist* 142 (1999): 71–76.

10. Bonnicksen, *op. cit.*

11. Suárez-Esteban, Alberto, Miguel Delibes, and José M. Fedriani. "Barriers or corridors? The overlooked role of unpaved roads in endozoochorous seed dispersal." *Journal of Applied Ecology* (2013). Available online. URL: https://besjournals.onlinelibrary.wiley.com/doi/full/10.1111/1365-2664.12080. Accessed June 27, 2023.

12. Cornett, James W. "Reading fan palms." *Natural History* 94 (1985): 64–73.

13. Bonnicksen, *op. cit.* Loeb, Robert E. "Evidence of prehistoric corn (*Zea mays*) and hickory (*Carya* spp.) planting in New York City: Vegetation history of Hunter Island, Bronx County, New York." *Journal of the Torrey Botanical Society* 125 (1998): 74–86.

14. Bonnicksen, *op. cit.*

15. Lanner, Ronald M. *The Piñon Pine: A Natural and Cultural History.* Reno: University of Nevada Press, 1981.

16. Betancourt, Julio L., William S. Schuster, Jeffry B. Mitton, and R. Scott Anderson. "Fossil and genetic history of a pinyon pine (*Pinus edulis*) isolate." *Ecology* 72 (1991): 1,685–1,697.

Notes 221

17. Fagan, Brian M. *Before California: An Archaeologist Looks at Our Earliest Inhabitants.* Lanham, MD: Altamira Press, 2004.

18. Schrader, James A., William R. Graves, Stanley A. Rice, and J. Phil Gibson. "Differences in shade tolerance help explain varying success in two sympatric *Alnus* species." *International Journal of Plant Sciences* 167 (5) (2006): 979–989. Gibson, J. Phil, Stanley A. Rice, and Clare M. Stucke. "Comparison of population genetic diversity between a rare, narrowly distributed species and a common, widespread species of *Alnus* (Betulaceae)." *American Journal of Botany* 95: 588–596 (2008). Schrader, James A., and William R. Graves. "Phenology and depth of cold acclimation in the three subspecies of *Alnus maritima*." *Journal of the American Society for Horticultural Science* 128 (2003): 330–336.

19. Rice, Stanley A., and J. Phil Gibson. "Is seedling establishment very rare in the Oklahoma seaside alder, *Alnus maritima* ssp. *oklahomensis*?" *Oklahoma Native Plant Record* 9 (2009): 59–63.

20. Janzen, Daniel H., and Paul S. Martin. "Neotropical anachronisms: The fruits the gomphotheres ate." *Science* 215 (1982): 19–27.

21. Bush, Leslie L. "Evidence for a long-distance trade in bois d'arc bows in 16th century Texas (*Maclura pomifera*, Moraceae)." *Journal of Texas Archeology and History* 1 (2014): 51–69. Available online. URL: https://scholarworks.sfasu.edu/ita/vol2014/iss1/76/. Accessed June 27, 2023.

22. Bonnicksen, *op. cit.*

23. Barlow, Connie. *The Ghosts of Evolution: Nonsensical Fruits, Missing Partners, and Other Ecological Anachronisms.* New York: Basic Books, 2000.

24. Warren, Robert J., II. "Ghosts of cultivation past—Native American dispersal legacy persists in tree distribution." *PLOS One* 11 (2016). Available online. URL: https://journals.plos.org/plosone/article?id=10.1371/journal.pone.0150707#pone.0150707.ref020. Accessed July 13, 2023.

CHAPTER 7

1. O'Fallon, Brendan D., and Lars Fehren-Schmitz. "Native Americans experienced a strong population bottleneck coincident with European contact." *Proceedings of the National Academy of Sciences* 108 (2011): 20,444–20,448. Available online. URL: https://www.pnas.org/doi/epdf/10.1073/pnas.1112563108. Accessed July 20, 2023. Thornton, Russell. *American Indian Holocaust and Survival: A Population History since 1492.* Norman: University of Oklahoma Press, 1990.

2. Littman, Robert J. "The Plague of Athens: Epidemiology and paleopathology." *Mount Sinai Journal of Medicine* 76 (2009): 456–467.

3. Wazer, Caroline. "The plagues that might have brought down the Roman Empire." *The Atlantic* (March 16, 2016). Available online. URL: https://www.theatlantic.com/science/archive/2016/03/plagues-roman-empire/473862/. Accessed April 9, 2024.

222   *Forgotten Landscapes*

4. *British Medical Journal.* "The plague that made England." *BMJ Opinion* (April 10, 2007). Available online. URL: https://blogs.bmj.com/bmj/2007/04/10/the -plague-that-made-england/. Accessed April 9, 2024.

5. Zinsser, Hans. *Rats, Lice, and History.* New York: Little, Brown, 1935. McNeill, William H. *Plagues and Peoples.* New York: Anchor, 1976. Aberth, John. *Plagues in World History.* Lanham, MD: Rowman & Littlefield, 2011.

6. Curtin, Philip D. *Death by Migration: Europe's Encounter with the Tropical World in the Nineteenth Century.* Cambridge, UK: Cambridge University Press, 1989. Curtin, Philip D. *Disease and Empire: The Health of European Troops in the Conquest of Africa.* Cambridge, UK: Cambridge University Press, 1998.

7. Hodgson, Dennis. "Malthus' Essay on Population and the American debate over slavery." *Comparative Studies in Society and History* 51 (2009): 742–770. Spengler, J. J. "Adam Smith on population." *Population Studies* 24 (1970): 377–388.

8. "The great dying 1616–1619, 'By God's visitation, a wonderful plague.'" Available online. URL: https://historicipswich.net/2023/11/17/the-great-dying/. Accessed April 9, 2024.

9. Mann, Charles C. *1493: Uncovering the New World Columbus Created.* New York: Knopf, 2011, p. 99.

10. Heckenberger, Michael J., et al. "Pre-Columbian urbanism, anthropogenic landscapes, and the future of the Amazon." *Science* 321 (2008): 1,214–1,217.

11. Faivre, Valentin. "Des complexes urbains de plus de 2,000 ans identifiés en Amazonie!" *Science et Vie* (April 2024), 32–33.

12. Schmidt, Morgan J., et al. "Intentional creation of carbon-rich dark earth soils in the Amazon." *Science Advances* 9 (2023). Available online. URL: https://www.science.org/doi/full/10.1126/sciadv.adh8499. Accessed April 9, 2024.

13. Fenner, Frank, ed. *Smallpox and Its Eradication.* Geneva, Switzerland: World Health Organization, 1989. Tucker, Jonathan B. *Scourge: The Once and Future Threat of Smallpox.* New York: Atlantic Monthly Press, 2001.

14. Alibek, Ken. *Biohazard: The Chilling True Story of the Largest Covert Biological Weapons Program in the World—Told from Inside by the Man Who Ran It.* New York: Delta, 1999.

15. Fenn, Elizabeth A. *Pox Americana: The Great Smallpox Epidemic of 1775–82.* New York: Hill and Wang, 2001.

16. Pearce-Duvet, Jessica M. C. "The origin of human pathogens: Evaluating the role of agriculture and domestic animals in the evolution of human disease." *Biological Reviews* 81 (2006): 369–382.

17. James Adair, 1775, quoted in Thornton, Russell. *The Cherokees: A Population History.* Lincoln: University of Nebraska Press, 1990.

18. Crosby, Alfred W. *Ecological Imperialism: The Biological Expansion of Europe, 900–1900.* Cambridge, UK: Cambridge University Press, 1986.

19. Dubos, René. *Man Adapting.* New Haven, CT: Yale University Press, 1965. Dubos, René. *Celebrations of Life.* New York: McGraw Hill, 1982.

20. Cox, David. "How Covid-19's symptoms have changed with each new variant." *BBC*, January 12, 2024. Available online. URL: https://www.bbc.com

/future/article/20240111-covid-19-how-does-its-symptoms-differ-from-flu. Accessed April 13, 2024.

21. Johnson, Stephen. *The Ghost Map: The Story of London's Most Terrifying Epidemic—and How It Changed Science, Cities, and the Modern World.* New York: Penguin, 2007.

22. Ewald, Paul. "Evolution of virulence." *Infectious Disease Clinics of North America* 18 (2004): 1–15.

23. Janeway, C. A., Jr., P. Travers, and M. Walport. "The generation of diversity in immunoglobulins." Chapter 4 of *Immunobiology: The Immune System in Health and Disease,* 5th ed. New York: Garland, 2001. Lindenau, Juliana Dal-Ri, Francisco Mauro Salzano, Ana Magdalena Hurtado, Kim R. Hill, et al. "Variability of innate immune system genes in Native American populations-relationship with history and epidemiology." *American Journal of Physical Anthropology* 159 (2016): 722–728.

24. Carrell, Jennifer Lee. *The Speckled Monster: A Historical Tale of Battling Smallpox.* New York: Penguin, 2004.

25. Gordon-Reed, Annette. *The Hemingses of Monticello: An American Family.* New York: Norton, 2008, 213.

26. Diamond, Jared. *Guns, Germs, and Steel: The Fates of Human Societies.* New York: Norton, 1997.

CHAPTER 8

1. Preston, William L. *Vanishing Landscape: Land and Life in the Tulare Lake Basin.* Berkeley: University of California Press, 1981.

2. Kosloff, Laura H. "Tragedy at Kesterson Reservoir: Death of a Wildlife Refuge Illustrates Failings of Water Law." *Environmental Law Reporter* (1985). Available online. URL: https://elr.info/sites/default/files/articles/15.10386.htm. Accessed May 3, 2024.

3. Bush, Evan. "A long-dormant lake has reappeared in California, bringing havoc along with it." *NBC News,* April 2, 2023. Available online. URL: https://www.nbcnews.com/science/environment/long-dormant-lake-reappeared-california-bringing-havoc-rcna75942. Accessed May 3, 2024.

4. Bourne, Joel K. "California's pipe dream." *National Geographic* (April 2010), 132–149.

5. Nash, Linda. *Inescapable Ecologies: A History of Environment, Disease, and Knowledge.* Berkeley: University of California Press, 2007.

6. Wright, Angus. *The Death of Ramón González: The Modern Agricultural Dilemma,* rev. ed. Austin: University of Texas Press, 2005.

7. Haslam, Gerald W. *The Other California: The Great Central Valley in Life and Letters.* Reno: University of Nevada Press, 1993. Johnson, Stephen, Gerald Haslam, and Robert Dawson. *The Great Central Valley: California's Heartland.* Berkeley: University of California Press, 1993.

224   *Forgotten Landscapes*

8. Viramontes, Helena Maria. *Under the Feet of Jesus*. New York: Plume, 1996.

9. Caltagirone, L. E., and R. L. Doutt. "The history of the Vedalia beetle importation to California and its impact on the development of biological control." *Annual Review of Entomology* 34 (1989): 1–16.

10. Carson, Rachel. *Silent Spring*. New York: Houghton Mifflin, 1962. Rice, Stanley A. "Resistance, evolution of." *Encyclopedia of Evolution*. New York: Facts on File, 2006.

11. Chia, Xing Kai, Tony Hadibarata, Risky Ayu Kristanti, Mohammed Noor Hazwan Jusoh, Inn Shi Tan, and Henry Chee Yew Foo. "The function of microbial enzymes in breaking down soil contaminated with pesticides: A review." *Bioprocess and Biosystems Engineering* 2024. Available online. URL: https://link .springer.com/article/10.1007/s00449-024-02978-6. Accessed May 3, 2024.

12. Wischemann, Trudy. "The role of land tenure in regional development: Arvin and Dinuba revisited." *California Geographical Society* 30 (1990): 25–51.

13. Earth Observatory, National Aeronautics and Space Administration. "San Joaquin Valley is still sinking." Available online. URL: https://earthobservatory .nasa.gov/images/89761/san-joaquin-valley-is-still-sinking#:~:text=Since%20the %201920s%2C%20excessive%20pumping,of%20California's%20San%20Joaquin %20Valley. Accessed May 3, 2024.

14. American Lung Association. "Most polluted cities." Available online. URL: https://www.lung.org/research/sota/city-rankings/most-polluted-cities. Accessed May 3, 2024.

15. Bonnicksen, Thomas M. *America's Ancient Forests: From the Ice Age to the Age of Discovery*. New York: John Wiley, 2000.

CHAPTER 9

1. Winnett, Robert. "President George Bush: 'Goodbye from the world's biggest polluter.'" *The Telegraph*, July 9, 2008. Available online. URL: https://www .telegraph.co.uk/news/worldnews/2277298/President-George-Bush-Goodbye -from-the-worlds-biggest-polluter.html. Accessed August 6, 2023.

2. Chua, Hannah Faye, Julie E. Boland, and Richard E. Nisbett. "Cultural variation in eye movements during scene perception." *Proceedings of the National Academy of Sciences USA* 102 (2005): 12,629–12,633.

3. Gall, Melanie. "Why insurance companies are pulling out of California and Florida, and how to fix some of the underlying problems." *The Conversation* (June 7, 2023). Available online. URL: https://theconversation.com/why-insur ance-companies-are-pulling-out-of-california-and-florida-and-how-to-fix-some -of-the-underlying-problems-207172. Accessed July 27, 2023. Henson, Bob. "A hail-battered June adds to billions in U.S. storm damage this year." *Yale Climate Connections* (June 29, 2023). Available online. URL: https://yaleclimateconnec

tions.org/2023/06/a-hail-battered-june-adds-to-billions-in-u-s-storm-damage -this-year/. Accessed July 27, 2023.

4. Rice, Stanley A. *Green Planet: How Plants Keep the Earth Alive*. New Brunswick, NJ: Rutgers University Press, 2008.

5. Marvel, Kate. "Lost cities and climate change." *Scientific American* (July 29, 2019). Available online. URL: https://blogs.scientificamerican.com/hot-planet /lost-cities-and-climate-change/. Accessed July 28, 2023.

6. Diamond, Jared. *The World until Yesterday*. New York: Penguin, 2012.

EPILOGUE

1. Demos, John. *The Unredeemed Captive: A Family Story from Early America*. New York: Knopf, 1994.

2. Blackmon, Richard D. *Dark and Bloody Ground: The American Revolution along the Southern Frontier*. Yardley, PA: Westholme Publishing, 2013.

3. Jacobs, Wilbur R. "The tip of the Iceberg: Pre-Columbian Indian demography and some implications for revisionism." *William and Mary Quarterly*, third series 31 (1974): 123–132. Jacobs, Wilbur R. *Dispossessing the American Indian: Indians and Whites on the Colonial Frontier*. Norman: University of Oklahoma Press, reprint, 1985. Stiffarm, Lenore A., and Phil Lane Jr. "The Demography of Native North America: A Question of American Indian Survival." Chapter 1 of Jaimes, M. Annette. *The State of Native America: Genocide, Colonization, and Resistance*. Boston: South End Press, 1992. Dorris, Michael. "Indians on the shelf," pp. 98–105 in Martin, Calvin, ed. *The American Indian and the Problem of History*. Oxford, UK: Oxford University Press, 1987.

4. Kennedy, John F. *A Nation of Immigrants*. New York: Harper & Row, 1964.

5. Fall, Thomas. *The Ordeal of Running Standing*. New York: Bantam, 1970. Verble, Margaret. *Stealing*. New York: HarperCollins, 2023.

6. Fitzsimmons, Tim. "Rick Santorum says 'there isn't much Native American culture in American culture.'" NBC News, April 26, 2021. Available online. URL: https://www.nbcnews.com/news/us-news/rick-santorum-says-there-isn-t-much -native-american-culture-n1265407. Accessed March 6, 2025.

7. Cronley, Connie. *A Life on Fire: Oklahoma's Kate Barnard*. Norman: University of Oklahoma Press, 2021, p. 153.

8. Blackburn, Bob, Duane King, and Neil Morton. *Cherokee Nation: A History of Survival, Self Determination, and Identity*. Tahlequah, OK: Cherokee Nation, 2019.

9. Payne, Russell. "'Excluding Indians': Trump admin questions Native Americans' birthright citizenship in court." *Salon*, January 25, 2025. Available online. URL: https://www.salon.com/2025/01/23/excluding-indians-admin-questions -native-americans-birthright-citizenship-in/. Accessed January 27, 2025.

226   *Forgotten Landscapes*

10. Board of Indian Commissioners, Annual Report, 1885, pages 90–91. Quoted in: Debo, Angie. *And Still the Waters Run: The Betrayal of the Five Civilized Tribes.* Princeton University Press, 1973 [1940], pages 21–22.

11. Debo, Angie. *And Still the Waters Run: The Betrayal of the Five Civilized Tribes.* Princeton, NJ: Princeton University Press, 1973 [1940]. Thorne, Tanis C. *The World's Richest Indian: The Scandal over Jackson Barnett's Oil Fortune.* Oxford, UK: Oxford University Press, 2005. Cronley, Connie. *A Life on Fire: Oklahoma's Kate Barnard.* Norman: University of Oklahoma Press, 2021.

12. Grann, David. *Killers of the Flower Moon: The Osage Murders and the Birth of the FBI.* New York: Vintage, 2017.

13. Inskeep, Steve. *Jacksonland: President Andrew Jackson, Cherokee Chief John Ross, and a Great American Land Grab.* New York: Penguin, 2015.

14. Wilms, Douglas C. "Cherokee land use in Georgia before Removal," pp. 1–28 in Anderson, William L. *Cherokee Removal: Before and After.* Athens: University of Georgia Press, 1991.

15. Parins, James W. *Literacy and Intellectual Life in the Cherokee Nation, 1820–1906.* Norman: University of Oklahoma Press, 2013.

16. Ehle, John. *Trail of Tears: The Rise and Fall of the Cherokee Nation.* New York: Anchor, 1988.

17. Perdue, Theda. *Cherokee Women: Gender and Culture Change 1700–1835.* Lincoln: University of Nebraska Press, 1998, p. 151.

18. Ro, Christine. "Why 'plant blindness' matters—and what you can do about it." BBC, February 24, 2022. Available online. URL: https://www.bbc.com/future/article/20190425-plant-blindness-what-we-lose-with-nature-deficit-disorder. Accessed May 6, 2024.

19. Wilson, Edward O. *Biophilia: The Human Bond with Other Species.* Cambridge, MA: Harvard University Press, 1986.

20. Wheeler, Graycen. "Stitt vetoes bill that would require Oklahoma irrigators to track how much groundwater they use." *Oklahoma Public Media Exchange,* May 3, 2024. Available online. URL: https://www.publicradiotulsa.org/local-regional/2024-05-03/stitt-vetoes-bill-that-would-require-oklahoma-irrigators-to-track-how-much-groundwater-they-use. Accessed May 6, 2024.

21. Witherspoon, Gary. *Language and Art in the Navajo Universe.* Ann Arbor: University of Michigan Press, 1977.

APPENDIX

1. Stewart, Omer C. *Forgotten Fires: Native Americans and the Transient Wilderness.* Norman: University of Oklahoma Press, 2002.

2. Bonnicksen, Thomas M. *America's Ancient Forests: From the Ice Age to the Age of Discovery.* New York: John Wiley, 2000.

3. Sherow, James E. *The Grasslands of the United States: An Environmental History.* Santa Barbara, CA: ABC-CLIO, 2007.

DEDICATION AND ACKNOWLEDGMENTS

I dedicate this book to the late professor Fakhri A. Bazzaz of Harvard University. He was a plant ecologist and my thesis advisor. When I started grad school, I thought science consisted of simple cause and effect. But Fakhri's studies wove together insights from many different academic fields. After studying with him, I could never again simply see a field or a forest; I saw them interconnected with the global ecosystem, humans and all of human history, and all of our current environmental problems. I have donated a portion of the proceeds of this book to the Dr. Fakhri and Dr. Maarib Bazzaz Plant Biology Fund at the University of Illinois.

I have not intended this book to be a scholarly treatment. Each chapter emerges from a different field of scholarly research, none of which corresponds exactly to my field of expertise (plant ecology). My purpose in writing this book was to open the eyes of interested readers to many aspects of Native ecology and history.

At first my interest in Native Americans was separate from my scientific interests. I grew up knowing but not having much appreciation of my Cherokee ancestry, of which ten generations are recorded. My high school history teacher Mike Green, and Wilbur Jacobs of the University of California at Santa Barbara, helped me see that the West was not won, but was lost, by conquest, not advancement. My detailed knowledge of my Cherokee ancestry was made possible by the tireless work of David Keith Hampton, who has documented the family history of the descendants of Nancy Ward, the last Beloved Woman of the Cherokee tribe, and my sixth great-grandmother. I knew about it, but only became inspired by Nancy Ward's story when I saw the musical *Nanyehi* by Becky Hobbs and Nick Sweet.

228   *Forgotten Landscapes*

I studied plant ecology for many years before I began to realize that the North American environment was not the product of white technology on a wilderness, but on a landscape that had been modified and constructed by Native Americans. This insight took me by surprise when I read books by Omer C. Stewart and Thomas M. Bonnicksen, and talked with recently retired Cherokee ethnobiologist Pat Gwin.

I am grateful to many other people as well. The writings of William Preston helped me re-see the San Joaquin Valley in which I had grown up and realize its significance long after I had moved away. I also learned to look closely at the natural world, especially plants, under the guidance of Reverend A. Luke Fritz, Presbyterian minister and Boy Scout leader; and botanists Bob Haller and Nancy Vivrette. My writing ability vastly improved under the guidance of Frank Darmstadt, former editor at Facts on File.

I am grateful to the reviewers of this book: Anna Vincent, director of the Spiro Mounds archaeological site in Oklahoma, who helped immensely with chapter 1; Jon Keeley, fire ecologist with the US Geological Survey, who helped immensely with chapter 2; Pat Gwin, Cherokee ethnobiologist, who helped with chapter 4; and William Preston, retired professor of social sciences at Cal Poly, San Luis Obispo, and expert on the San Joaquin Valley, who helped with chapter 8 and parts of chapter 3. Any errors of fact are my own. I also wish to thank my agent, Rita Rosenkranz, and the editors at Globe Pequot, especially Melissa Hayes, whose attention to detail was amazing.

INDEX

Note: Photos are indicated by italicized page references.

acacias, 131
acorns, 134, *135*
acorn woodpeckers, 47, *48*
Adair, James, 149–50
Africa: agroforestry in, 94; Asia and,
 15, 151, 157, 159; cultures of, 143;
 Europe and, 76, 145; farms in,
 114; Mesopotamia, 112–15; North
 America and, 33–34; World Health
 Organization in, 147–48
agriculture: artificial crops, 169–70;
 biotechnology in, 85–86; by
 Cherokees, 83–84; in Europe, 87,
 123; history of, 23; hunting and,
 20–21; in Imperial Valley, 130;
 in Mexico, 106; in Mississippian
 civilization, 11; in modernity,
 176–77; monocultures in, 85–87,
 86; Native Americans and, 82–85,
 89; in North America, 112; in
 Oklahoma, 91–93, *92*, 181–82;
 polycultures and, 88–97, *91–92*;
 before settlers, 18; shifting, 96–97;
 in South America, 11; wild plants
 in, 101. *See also* farms
agroforestry, 88–97, *91–92*, 177–78
air quality, 168–69, *169*

Akimel O'odham tribe, 110, 112
Alaska, 119
Algonquin tribe, 73
alligators, 69
Amazon rainforest, 146–47
American bison, 7, 41, 45–46, 50, 64,
 73
American Progress, 185, *185*
America's Ancient Forests (Bonnicksen),
 42
"America the Beautiful," 140–41, 183
Anasazi civilization: evidence of,
 102–3, *105*, *107*, *109*; irrigation
 by, 110–12; scholarship on, 101–8,
 115, 119
Anderson, M. Kat, 42, 54–55
Anikutani, 28
animals. *See specific animals*
Antarctica, 33–34
anthropology, 54–56, 73–74, 136–37
antibodies, 155–56
Anza-Borrego Desert State Park, 130,
 130
Apaches, 101–2
Apess, William, 123
archaeology: by accident, 30;
 Archaeological Resources

229

230   *Forgotten Landscapes*

Protection Act, 32; botany and, 18; from Chaco Canyon, 102; DNA evidence in, 136–37; evidence from, 11; from Native Americans, 30–31; in North America, 15–17; in Oklahoma, 36; pollen in, 52–53; runic carvings, 25; salinization in, 113–14; science and, 3; with tree rings, 41–42, *103*, 103–4

Armendáriz, Pedro, 190

Army Corps of Engineers, 118, 165

artificial crops, 169–70

artificial ecology, 172

artificial warmth, 170–72, *171*

Asia: Africa and, 15, 151, 157, 159; cultures of, 143; Europe and, 10, 12, 17; history, 76; World Health Organization in, 147–48

As Long as the Rivers Run (Slade), 195

At Play in the Fields of the Lord (Matthiessen), 151

Attakullakulla, 28

Audubon, John James, 62

The Awakening Land (Richter), 184

Aztec Empire, 10, 12–13

Babylonia, 12, 113, 114–15

bacterium, 154–55, 174–75

balanced pathogenicity, 151–54

Bambi (film), 33

barges, 20

Barnes, Carl, 91–93

bartering, 20–21

Bartram, William, 18, 45, 69

Bates, Katharine Lee, 140, 183

bathing, 22–23

beans. *See* Three Sisters legend

bees, 126

berries, 80

Beverley, Robert, 67

biodiversity, 173

biology, 55, 66, 71–72

biotechnology, 85–86

bison, 7, 41, 45–46, 50, 64, 67–69, 73

blackberries, 127–28

black walnut trees, *125*

Bloomberg, Michael, 79

bois d'arc trees, 138

Bolivia, 12

Bonnicksen, Thomas M., 42, 54, 132, 139, 201, *201–7*

Boone, Daniel, 184

botany, 18, 45, 69

Boudinot, Elias, 195, 197

Boulding, Kenneth, 175–76

Bregman, Rutger, 9–10

British Empire, 144, 157, 186

Bryant, Edwin, 74

bubonic plague, 144–45

Bullock, William, 47

Bureau of Biological Survey, 66, 71–72

Bureau of Reclamation, 118, 165

Burnaby, Andrew, 73

Bush, George H. W., 174

Bush, George W., 174

Bushyhead, Jesse, 193

Cabeza de Vaca, Álvar Núñez, 44, 49

Cahokia, 12–13, *14*, 15–17, 21, 30, 88, 106–7

California Aqueduct, 118, *118*

California Division of Forestry, 54

California Natives: Cornett on, 131; forests and, 34, *35*, 45; Great Lakes Natives and, 93–94; grizzly bears and, 52; history of, 6, 56–57; in Imperial Valley, 130; La Brea Tar Pits and, 78; scholarship on, 47–49, *48*, 50, 73; with wildflowers, 56–57; wtih environment, 43–44

Canada, 24–25, 40, 52–53, 62

Carson, Kit, 127

Carson, Rachel, 70, 167

Catholic Church, 144

Catlin, George, 44, 51–52

cattle, 165
Cavelier, René-Robert, 73
Centers for Disease Control, 191
Central America, 12, 90–91
Central Valley, 117–19, 161–63. *See
also* San Joaquin Valley
ceremonial drinks, 137–38
Chaco Canyon, 101–8, *102–3*, *105*,
107, *109*, 109–10
Champlain, Samuel de, 74
Channel Islands, 11
Chardon, Francis, 150
Cherokees: agriculture by, 83–84;
Anikutani for, 28; census of, *194*;
Christianity for, 5; cool shade and,
95; descendants of, 191, 193; as
diplomats, 159; disease for, 75;
with dogs, 16; farms by, 50–51;
government of, 195–96; Green
Corn Festival for, 22–23; history of,
189; hunting by, 65; for Jefferson,
29; language of, 197; legacy of, 173,
181; in migration, 137; Muskogee
tribe and, 128; North America
for, 3, 5; racism against, 193–94;
scholarship on, 53; seed dispersal
by, 140; Sequoyah on, 57; sign
language by, 181; smallpox and,
149–50; stereotypes of, 6; Three
Sisters legend and, 89–93, *91–92*;
trade with, 66; on Trail of Tears, 3,
6, 59–60, 121, 127; in war, 27–28;
Ward for, 1–2, 6–7, 137–38;
women, 5; Women's Council for,
28
chestnuts, 132
Chickamauga communities, 28, 150
chickenpox, 151
Chile, 11
China, 22, 143
cholera, 154
Christianity: for Cherokees, 5; farms
in, 96–97; Jesus in, 197–99;

King James Bible, 22; from Old
Testament, 97, 114–15, 185;
plagues in, 70
Cincinnati Zoo, 63
citrus trees, *163*, 164, 169–72, *171*
civilization, 12–13, 24–26
Civil War, 57
Clark, Galen, 60
clean cities, 21–23, 32
Clements, Forrest, 31
climate change, 178–80
Clovis tips, 11
coastal grasslands, 56–57
coffeetrees, 139
Collapse (Diamond), 13
colonialism, 1, 10, 145–46, 175
Colorado River, 101–2, 116–17, 127
Columbus, Christopher, 6, 9–10,
24–25, 153, 184–85
Comanche tribes, 83
communal hunting, 11
communication, 12, 51, 180–81, 187,
197
*A Connecticut Yankee in King Arthur's
Court* (Twain), 22
cool shade, 94–95
Cornett, James W., 131
cornfields, 85, *86*, 96
Coronado, Francisco de, 68
Cortés, Hernán, 10
COVID-19, 152–55
cowpox, 149
coyotes, 128–29
Craig Mound, 31, *32*
Crespí, Juan, 48–49
crop rotation, 96–97
Crosby, Alfred, 151
crows, 96
cultural diversity, 180–81
Cultural Element Distribution project,
54
Cuming, Alexander, 28
Curtis, Tony, 25

232   *Forgotten Landscapes*

Dakota tribe, 68, 175–76
Darwin, Charles, 64, 145
dates, 129–32, *130*
Davidson, George, 50
Dawes, Henry L., 189
DDT, 167–68
deciduous forests, 34, 36
deer/venison, 65–67, 73, 81, 106
deforestation, 108
Delmarva region, 136–37
democracy, 2, 29
dendrochronology, 41–42
desert willow tree, 131
De Smet, Pierre-Jean, 74
Diamond, Jared, 13, 159
diplomacy, 27–28
discovery, 196–97
disease: bathing and, 22–23; bubonic
    plague, 144–45; for Cherokees, 75;
    COVID-19, 152–55; from Europe,
    21, 61, 65, 107–8, 129, 143–46,
    186; from fungus, 23–24; immune
    systems and, 155–59; immunization
    from, 147–48; influenza, 151–53;
    inoculation from, 157; for Native
    Americans, 150–55; plagues, 23;
    prevention, 89; settlers and, 6,
    53–54; smallpox, 69, 146–50, *147*,
    153–54, 156–59; tuberculosis, 144;
    vaccinations, 152–53; World Health
    Organization on, 147–48
Disney, 33
DNA evidence, 136–37
dogs, 15–16
Donck, Adriaen van der, 44, 45–46
Donner Party, 164–65
Douglas, Kirk, 25
Douglas firs, 56, 104
Dragging Canoe, 6
drip irrigation, 115–16
droughts, 108
Dubos, René, 152
dumping, 190–91
Dust Bowl, 141

Ecological Imperialism (Crosby), 151
ecology, 42
Ecuador, 12
edible fruits, 127–32, *130*
edible nuts, 132–34, *135*
Egypt, 12–13, 102, 114, 129
elephant trees, 131
Emerald Forest (film), 147
Emerson, Ralph Waldo, 84–85
Emmons, George, 45
English walnut trees, *125*
the Enlightenment, 29
environment: biodiversity of,
    173; California Natives with,
    43–44; coastal grasslands, 56–57;
    environmental biology, 55;
    environmentalism, 74–75, 117;
    fire in, 33–34, *35–39*, 36–39;
    for government, 7; grasslands,
    37; habitat management, 38, *38*,
    45, 51; Native Americans with,
    33–34, 41–45, 59–60, 174–77; in
    New Deal, 54; of North America,
    121, *122*, 123; oil from, 20;
    pastures, 50–51; philosophy and,
    183–86, *185*, 196–99; Pleistocene
    extinction, 76–78; prairies, 33–34;
    for settlers, 46; soil erosion, 88–89;
    soil fertility, 46; of Spiro, 121, *122*;
    sustainability in, 77–78; tundra,
    40; undergrowth, 46–47; wildfires
    in, 39; wildlife management,
    74–75. *See also* agriculture; forests;
    irrigation
Environmental Protection Agency, 191
Erdrich, Louise, 197
ergot poisoning, 23–24
Erriksson, Leif, 26
escaped fires, 51
ethnography, 43
Etowah Mound, 17, *17*, 30
Europe: Africa and, 76, 145;
    agriculture in, 87, 123; Asia and,
    10, 12, 17; Aztec Empire and,

12; colonialism by, 1; Columbus and, 6; communication in, 51; culture of, 28; disease from, 21, 61, 65, 107–8, 129, 143–46, 186; the Enlightenment in, 29; explorers from, 72; history of, 40; immigration from, 188; Jamestown colony from, 83; medieval, 9–10; Native Americans and, 2, 12–13, 18, 21–26, 79; North America and, 21–23, 44, 81, 132, 187–88; religion in, 22; Roman Empire in, 27; scholarship on, 151; scurvy in, 23; settlers from, 32, 43, 74–75, 127, 176–77, 183–86, *185*; slavery in, 187; smallpox in, 69, 157–59; South America and, 15; toilets in, 21–22; tribes in, 6; Vikings in, 25; winters in, 23–24
Ewald, Paul, 154–55
explorers, 201, *201–7*
extinction, 76–77

Fall, Thomas, 187
fallow, 96–97
"Farming" (Emerson), 84–85
farms: in Africa, 114; agroforestry and, 88–97, *91–92*; in Cahokia, 88; cattle, 165; in Christianity, 96–97; cool shade on, 94–95; cornfields, 85, *86*; crop rotation, 96–97; drip irrigation on, 115–16; forests and, 79–83, *82*, 97; history of, 83–85; Hohokam tribes with, 178; in Mexico, 81; in Middle East, 87; Native Americans and, 95–96, 141; pasture on, 50–51; soil erosion on, 177; square, 86–87; in United States, 165–66
Featherstonhaugh, George W., 53
Federal Writers' Project, 118
Fenner, Frank, 147–48
fire: in coastal grasslands, 56–57; communication with, 51; in

environment, 33–34, *35–39*, 36–39; forest fires, 181–82; forests after, 37–38; in giant sequoia forests, 57–59, *58–59*; history of, 41–45, 54–56, 77–78, 201, *201–7*; for hunting, 49–50; Native Americans and, 33, 41, 45–46, 52–53, 132; in North America, 40; on pastures, 50–51; pollution from, 59–60; safety with, 51–52; for soil fertility, 46; suppression, 41; as tool, 72–74; undergrowth and, 46–47; wildfires, 39; wild foods and, 47–49, *48*
Fish and Wildlife Service, 71–72
fishing, 72–74
Fleischer, Richard, 25
food rationing, 178
forests: agroforestry, 88–97, *91–92*; Bonnicksen on, 139; California Division of Forestry, 54; for California Natives, 34, *35*, 45; in Canada, 52–53; deciduous, 34, 36; deforestation, 108; Douglas firs, 56, 104; ecology of, 42; farms and, 79–83, *82*, 97; after fire, 37–38; forest fires, 181–82; giant sequoia forests, 57–59, *58–59*; hardwood forest, *122*; National Forest Service, 39; Native Americans and, 41, 46–47; in North America, 40; in Oklahoma, 38, *38*; orchards compared to, 124, *125*, 126; pollinators in, 126; of San Joaquin Valley, 134; scholarship on, 45–46; science of, 44–45; tree rings, 41–42, 103, *103*, 103–4; tree trunks, 57–58; tree types, 56–59, *58–59*; unburned, *37*, 39, *39*
Forgotten Fires (Lewis and Anderson), 42
Founding Fathers, 2, 29
France, 28, 153, 158
Franklin, Benjamin, 29
Friant-Kern Canal, *162*

234   *Forgotten Landscapes*

fruit, 127–32, *130*
Fuegians, 100
fungus, 23–24, 132

game pieces, 139
game sticks, 139–40
Gast, John, 185, *185*
gathering food, 47–49, *48*, 50–51, 80, 83–85
genocide, 186–87
geology, 53
giant sequoia forests, 57–59, *58–59*
global climate change, 178–80
Glome Dahl, 25
gold, 6
Gold, Harriet, 195
Grant, James, 84
grapes, 127–28
grasslands, 37
Great Depression, 117–18
Great Lakes Natives, 93–94
Greece, 29
Green Corn Festival, 22–23
Greenland, 33–34
grizzly bears, 52
groves. *See* orchards
gutters, 21
Gwin, Pat, 91–93

habitat management, 38, *38*, 45, 51
Hamar, Ralph, 47
hardwood forest, *122*
Hayward, Susan, 190
health, in North America, 23–24
Heavener, 25–26
height, 210n18
Hemings, Sally, 158
Hennepin, Louis, 73
Hicks, Edd, 51, 96, 189
Hilderbrand, Nannie, 195
Hobbs, Becky, 5, 176–77
Hohokam tribes, 110–13, 115–16, 119, 178, 187

Holland, 186
holly, 137–38
Holt, Catharine, 73–74
honey locusts, 139–40
horses, 67–68, 138, 175–76
Hospital Rock, *135*
humanity: balanced pathogenicity and, 151–52; history of, 9–11, 33–34, *35–39*, 36–39, 150–51; immune systems, 155–59; in Little Ice Age, 59–60
human waste, 21–22
hunting: agriculture and, 20–21; American Bison, 67–69; capturing prey, 74; communal, 11; deer/venison, 65–67, 73; extinction and, 76–77; fire for, 49–50; fishing and, 72–74; gathering and, 83–85; by Native Americans, 15–16, 77–78; passenger pigeons, 61–65, 74–75; resurgence after, 69–72; wildlife management and, 74–75

ice ages, 40, 59–60, 77
immigration, 184, 187–88
immune systems, 155–59
immunization, 147–48
Imperial Valley, 130
Incas, 12
India, 143
Indian Health Service, 191
Indian Removal Act, 193
Indian Territory, 6–7, 31, 50–51
influenza, 151–53
inoculation, 157
Inuits, 99–100
invasive species, 182
Iroquois Confederacy, 29
irrigation: by Anasazi civilization, 110–12; in Chaco Canyon, 101–8, *102–3*, *105*, *107*, *109*, 109–10; history of, 119; by Hohokam tribes, 187; by Native Americans, 71, 87;

in North America, 99–100; with Paiute canals, *100*, 100–101; power for, 115–18, *118*; salinization and, 112–15; in San Joaquin Valley, 71, 115–16, 168
Israel, 97, 115–16

Jackson, Andrew, 140
Jacobsen, Thorkild, 113–14
James (king), 145–46
Jamestown colony, 83
Japan, 164
Jefferson, Thomas, 29, 158
Jesus Christ, 197–99
Johnson, E., 46

Kaibab Plateau, 67
kanuchi paste, 121
Kennedy, John F., 187–88
Kentucky coffeetree, 139
Kesterson National Wildlife Refuge, 166
Killers of the Flower Moon (film), 190
King, E. Sterling, 195
King, James, 52–53
King James Bible, 22
knowledge, 12–13
Kubla Khan, 123–24

La Brea Tar Pits, 78
land allotment, 189–90
land management, 7
Land Ordinance (1785), 86–87
Landrum, E. M., 188
language, 12, 187, 197
La Salle, Sieur de, 73
The Last Report on the Miracles at Little No Horse (Erdrich), 197
legal maneuverings, 189–90
Leopold, Aldo, 66–67, 75
Lewis, Henry T., 42, 54–55
lice, 155
Lindstrom, P., 44

Little Ice Age, 59–60
livestock, 66
Lodge, Henry Cabot, 188
lysergic acid diethylamide, 23–24

Maathai, Wangari, 94
Madagascar, 76
Maidu tribe, 73
maize. *See* Three Sisters legend
malaria, 155
Malathion, 167
Malthus, Thomas, 145
Mandan tribe, 150
Mantle Rock, 1, 3, *4*, 41, 195–97
Martin, Joseph, 195
Mather, Cotton, 62
Matthiessen, Peter, 151
Mayans, 12, 17
medicinal bark, 136–37
medieval Europe, 9–10
Mesa Verde, 102, 108, *109*, 110
Mesopotamia, 112–15
Mexico: agriculture in, 106; Aztec Empire in, 10; Central America and, 90–91; farms in, 81; history of, 15; Native Americans in, 146, *147*; before settlers, 11; Spain and, 165; Three Sisters legend in, 110–11, 119; trade in, 82; United States and, 167
Middle Ages, 27
Middle East, 12, 82, 87, 97, 129
migration: Cherokees in, 137; to Indian Territory, 50–51; by Native Americans, 49–50, 111–12; on Trail of Tears, 128; in Younger Dryas era, 179–80
Miller, Joaquin, 60
mining, 190–91, *193*
missionaries, 48–49
Mississippian civilization: agriculture in, 11; barges in, 20; Cahokia for, 13; collapse of, 180; history of,

236 *Forgotten Landscapes*

10, 16–17, 30–32, *32*, 187; Native Americans in, 27–32, *32*; pecans and, 121, *122*, 123; sacrifice in, 15
Mississippi River drainage system, 6
modernity: agriculture in, 176–77; fire suppression in, 41; global climate change in, 178–80; government in, 41; Mantle Rock in, 3, *4*; modern orchards, 123–24, *125*, 126, 140–41, *141*; Native Americans and, 173, 181–82; North America in, 196–99; science in, 165; smallpox in, 147–48; Spiro in, 30–32, *32*
Mohave Desert, 34
Mongolia, 33–34
Monk's Mound. *See* Cahokia
monocultures, 85–87, *86*
Montagu, Mary Wortley, 157
Monte Verde, 11
Montgomery, Archibald, 83
Mooney, James, 47, 84, 139
Moore's Law, 175
More, Thomas, 9
Mormon locust plague, 70
Morton, Thomas, 46–47
mosquitoes, 155, 166
mouse plague, 71–72
Moytoy, 28
Muir, John, 117
mulberries, 127–28
Mulholland, William, 117
Muskogee tribe, 20, 128, 189–90

Nabataeans, 115
Nanyehi. *See* Ward, Nancy
Nanyehi (musical), 5
National Forest Service, 39
National Park Services, 22
A Nation of Immigrants (Kennedy), 187
Native Americans. *See specific topics*
native horses, 138
Navajo tribe, 101–2, 116, 127

New Deal, 31, 54, 117–18
Noem, Kristi, 79
non-food products, 50
North America: Africa and, 33–34; agriculture and, 112; archaeology in, 15–17; Cahokia and, 12–13, *14*, 15–16; Cherokees and, 3, 5; clean cities in, 21–23; colonialism in, 175; before Columbus, 9–10; droughts in, 108; edible fruits in, 127–32, *130*; edible nuts in, 132–34, *135*; environment of, 121, *122*, 123; Europe and, 21–23, 44, 81, 132, 187–88; fire in, 40; forests in, 40; health in, 23–24; history of, 2, 10–11, 41, 159, 175–76, 183–86, *185*; irrigation in, 99–100; Little Ice Age in, 59–60; medicinal bark in, 136–37; Mississippian civilization in, 27–32, *32*; in modernity, 196–99; Native Americans and, 42–43, 80–81, 179; Old Testament and, 184–85; settlers and, 24–26, 63, 140–41, *141*, 181–82; South America and, 91–92; Spiro and, 16–18, *17*; trade networks in, 18, *19*, 20–21; Vikings in, 24–25; for Ward, 6; wolves in, 66. *See also specific topics*
nutrition, 24
nuts, 132–34, *135*

Ogallala Aquifer, 141
oil, 20, 189–90
Oklahoma: agriculture in, 91–93, *92*, 181–82; archaeology in, 36; forests in, 38, *38*; government of, 1, 188, 196–97; Native Americans in, 6–7; oil in, 189–90; Pocola Mining Company in, 30; before settlers, 16–18, *17*; Southeastern Oklahoma State University, 25; Spiro Mound in, 30–32, *32*; Turkey Mountain

Urban Wilderness in, 20; University of Oklahoma, 31; WPA in, 31

Old Testament, 97, 114–15, 184–85

Old World. *See* Europe

"On Jordan's Stormy Banks I Stand" (Stennett), 184–85

"Operation Warp Speed," 155

orange blossoms, *161*

orange groves, 164, 169–72, *171*

orchards: acorns and, 134, *135*; alders in, 136–37; with bois d'arc trees, 138; dates in, 129–32, *130*; food from, 139–40; with holly, 137–38; modern, 123–24, *125*, 126, 140–41, *141*; mulberries in, 127–28; of Native Americans, 121, *122*, 123, 127; pawpaws in, 128–29; pine nuts in, 132–34; in San Joaquin Valley, 86, 123–24

The Ordeal of Running Standing (Fall), 187

ore mines, 190–91, *193*

The Origin of Species (Darwin), 64

Osage tribe, 190

Ostenaco, 28–29

Owens Valley, *100*, 100–101, 117, 119

Owl Canyon, 134

Paiute canals, *100*, 100–101

Paleoindians. *See specific topics*

Palestine, 97

Palo verde, 131

parasites, 151–52, 157

Parker, Cynthia Ann, 26

passenger pigeons, 61–65, 74–75, 129

pastures, 50–51

pawpaws, 128–29

pecan trees, 121, *122*, 123, 127

Pequot tribe, 186–87

persimmons, 128–29

Peru, 12

pesticides, 70, 167–68

Petra, 115

Pettit, Elizabeth Hilderbrand: exile of, 191, *192*, 193–95, *194*; life of, 1, 3, 41, 90, 121

Pettit, James, 195

Pettit, Minerva, 193

Philip (king), 123

pigeons, 61–65

pine nuts, 132–34

Pinkney, Edward Coote, 195

piñon pine trees, 132–33

Plague of Athens, 144

plagues, 23, 70–72

Plains Natives, 67–69

plant distribution, 127, 140, *141*

Pleistocene extinction, 11, 76–78

Pocola Mining Company, 30–31

pollen, 52–53, 126

pollution, 59–60, 168–69, *169*, 178–80

Polo, Marco, 22

polycultures, 88–97, *91–92*, 177

power, for irrigation, 115–18, *118*

prairies, 33–34, 40, 44, 49–50, 52–53

Preston, Robert, 190

Pueblo tribe, 110

Quapaw Natives, 191

quarantines, 158

raccoon, 128–29

racial identity, 194–95

racism, 188–90, 193–94

Rats, Lice, and History (Zinsser), 144

Rawlings, Marjorie Kinnan, 16

"Red Book" (Fenner), 147–48

Reisner, Marc, 118

religion: for Anikutani, 28; in British Empire, 186; Catholic Church and, 144; in Europe, 22; history of, 114, 197–98; King James on, 145–46; in Middle East, 97; missionaries, 48–49; of Native Americans, 16; Old Testament, 97, 114–15, 185;

238   *Forgotten Landscapes*

philosophy of, 186–87; songs of, 184–85. *See also* Christianity
resurgence, after hunting, 69–72
Revere, Paul, 51
Revolutionary War, 51, 84, 148, 158, 187
Richter, Conrad, 184
Riddle, George, 47–48
Ridge, John, 5
Rocky Mountain Natives, 74
Roman Empire, 27, 144
Roosevelt, Franklin, 117–18
Ross, John, 1, 193
runic carvings, 25
Rutherford, Griffith, 84
rye plants, 23–24

safety, with fires, 51–52
Sagard-Thépdat, Gabriel, 62
salinization, 112–16
sandstone, 104, *105*
San Joaquin Valley: air quality in, 168–69, *169*; artificial crops in, 169–70; artificial warmth in, 170–72, *171*; California Aqueduct in, 118, *118*; forests of, 134; irrigation in, 71, 115–16, 168; Native Americans in, 164–65, 172; orchards in, 86, 123–24; pesticides in, 167–68; settlers in, *161–63*, 161–66
Santorum, Rick, 188
satiation, 133
Scandinavia, 144
Scorsese, Martin, 190
scurvy, 23
seaside alder trees, 136–37
seed dispersal, 127–29, 136–37, 140
seed transportation, 138
Sequoia National Park, 57, *135*
Sequoyah, 5, 57
settlers: agriculture before, 18; arrogance of, 187–88; in colonialism, 145–46; disease and,

6, 53–54; environment for, 46; from Europe, 32, 43, 74–75, 127, 176–77, 183–86, *185*; in Jamestown colony, 83; legacy of, 173, 181–82; Mexico before, 11; with mulberries, 128; Native Americans and, 5, 44–45, 97, 143–46, 155–59, 201, *201–7*; North America and, 24–26, 63, 140–41, *141*, 181–82; oil and, 20; Oklahoma before, 16–18, *17*; in San Joaquin Valley, *161–63*, 161–66; from Spain, 68, 107–8; trade with, 66; treaties with, 6; violence by, 143
sexism, 5
Sherow, James E., 201, *201–7*
shifting agriculture, 96–97
shrublands, 49
Siberia, 133–34
sign language, 180–81
Silent Spring (Carson, R.), 70
Sioux tribes, 6
Six Nations, 29
skraelings (miserable people), 25–26
Slade, Sam J., 195
slavery, 187
smallpox: balanced pathogenicity with, 153–54; in Europe, 69, 157–59; in modernity, 147–48; for Native Americans, 146–50, *147*, 156; in Revolutionary War, 158; in United States, 148–49
Smith, Adam, 145
Smith, John, 47, 128
smoke trees, 131
Smythe, William, 110
Snow, John, 154
soil epidemics, 158–59
soil erosion, 88–89, 177
soil fertility, 46
Soto, Hernando de, 128
South America: agriculture in, 11; artifacts in, 11; Central America

Index   239

and, 12; Europe and, 15; Fuegians in, 100; North America and, 91–92
Southeastern Oklahoma State University, 25
Spain, 68, 107–8, 123, 153, 165, 176
Spiro: Cahokia and, 16–17, 21, 30, 106–7; environment of, 121, *122*; Heavener and, 25; in modernity, 30–32, *32*; North America and, 16–18, *17*; pecans at, 127; Vikings in, 26. *See also specific topics*
square farms, 86–87
squash. *See* Three Sisters legend
Stealing (Verble), 187
Stennett, Samuel, 184–85
Sternberg, George M., 44–45
Steward, Julian, 54
Stewart, Omer Call, 42–43, 54–55, 201, *201–7*
sugar maple, 139
sumpweed, 81, *82*
sustainability, 77–78
Sweet, Nick, 5, 176–77
syphilis, 21

Tar Creek Superfund Site, 191
tarweeds, 47–48
Thoreau, Henry David, 34
Thornton, Russell, 84
Three Sisters legend, 89–93, *91–92*, 110–11, 119
ticks, 155
Timberlake, Henry, 28
Tiwanaku, 12
tobacco, 93
Tohono O'odham tribe, 110, 112
Tolkein, J. R. R., 34, 36
toxic plants, 80–81, 85–86
trade networks, 18, *19*, 20–21, 32, 82
Trail of Tears: Cherokees on, 3, 6, 59–60, 121, 127; history of, 7, 191, 193; migration on, 128; Muskogee

tribe before, 20; policy for, 1; for Van Buren, 84–85
transportation, 18, *19*, 20
treaties, 6
tree rings, 41–42, 103, *103*, 103–4
trees. *See specific trees*
tree trunks, 57–58
Trump, Donald, 155
Tsiyu Gansini (Dragging Canoe), 6, 27–28, 150, 176–77
tuberculosis, 144
Tulsa (film), 190
tundra, 40, 55, 99–100
Turkey, 157
Turkey Mountain, 18, *19*, 20, 25
Twain, Mark, 22

unburned forests, *37*, 39, *39*
undergrowth, 40, 46–47, 51–52, 97
uniformity, 85
United Nations, 147–48, 174
United States: ancient ruins in, 13, *14*, 15; Anza-Borrego Desert State Park, 130, *130*; Archaeological Resources Protection Act in, 32; army, 1; Army Corps of Engineers, 118, 165; bison in, 7, 41, 45–46, 50, 64, 73; Bureau of Biological Survey, 66, 71–72; Bureau of Reclamation, 118, 165; Civil War, 57; climate change in, 178–79; Colorado River, 101–2, 116–17, 127; cornfields in, 85, *86*; culture, 33; Delmarva region of, 136–37; diplomacy with, 27–28; farms in, 165–66; Fish and Wildlife Service, 71–72; Founding Fathers, 2; France and, 158; gold in, 6; Great Depression in, 117–18; history, 61–62; immigration in, 184; Imperial Valley of, 130; Indian Health Service in, 191;

240   *Forgotten Landscapes*

Indian Removal Act, 193; Indian Territory in, 6–7; Kaibab Plateau in, 67; Kesterson National Wildlife Refuge, 166; land allotment in, 189–90; land management in, 7; Land Ordinance, 86–87; Mexico and, 167; National Forest Service in, 39; National Park Services in, 22; Native Americans and, 196–99; New Deal in, 31; "Operation Warp Speed," 155; ore mines in, 190–91, *193*; Owl Canyon, 134; policy, 20, 32, 68–69, 191, *192*, 193–95, *194*; poverty in, 9; prairies in, 33–34, 44; racism in, 188–90; Revolutionary War, 51, 148, 158; Sequoia National Park, 57, *135*; smallpox in, 148–49; Turkey Mountain in, 18, *19*, 20, 25

University of Oklahoma, 31

Utopia (More), 9

Utopia for Realists (Bregman), 9–10

vaccinations, 152–53

Van Buren, Martin, 84–85

Vancouver, George, 148

variola minor, 154

venison, 65–67

Verble, Margaret, 187

Verrazzano, Giovanni da, 44

Vikings, 24–26, 210n18

The Vikings (film), 25

viruses, 151–52

Waite, Buck, 195

Wallace, Alfred Russel, 145

walnut trees, *125*, 170

war, 10, 27–28, 84. *See also specific wars*

Ward, Betsy, 195

Ward, Bryant, 194–95

Ward, Nancy: for Cherokees, 1–2, 6–7, 137–38; descendants of,

194–95; leadership of, 5, 191, 193; legacy of, 176–77; North America for, 6

Washington, George, 158

water. *See* irrigation

water technology, 99–100, 115–19, *118*

weeds, 174–75

Wells, R. W., 47, 50

Whealy, Kent, 93

white blood cells, 155–56

white eagle seeds, 91–93, *92*

wildfires, 39

wildflowers, 34, *36*, 56–57

wild foods, 47–49, *48*

wildlife, 62, 71–72. *See also specific wildlife*

wildlife management, 74–75

wild plants, 101

The Wild Rose of Cherokee (King, E.), 195

William and Mary College, 28–29

Williams, Eunice, 26

Williamson, Andrew, 84

wind machines, *171*

wolves, 66–67, 75

women, 5, 28, 96

wood, 138

Wood, William, 46–47

Works Progress Administration (WPA), 31

World Health Organization, 147–48, 152

WPA. *See* Works Progress Administration

Wright, C. W., 134

The Yearling (Rawlings), 16

Yokuts tribe, 165

Younger Dryas period, 179–80

Zinsser, Hans, 144

Zuni tribe, 110